Financial Mathematics

Key Concepts and Tools for SOA Exam FM & CAS Exam 2

Olivier Le Courtois, PhD, FSA, CFA, CERA, FRM

Financial Mathematics.
Key Concepts and Tools for SOA Exam FM & CAS Exam 2.

Copyright © 2019 by Olivier Le Courtois.

All rights reserved. No part of this book may be reproduced or transmitted in any form or by any means, electronic or mechanical, including photocopying, recording, or by any information storage and retrieval system without the written permission of the publisher, except where permitted by law.

Cover design by Emiko Muraoka.

Contents

1	**Introduction**	**1**
2	**Time Value of Money**	**5**
	2.1 Interest Rates	5
	2.2 How to Deal with Sub-Periods	9
	2.3 Continuous Compounding	12
	2.4 Pricing	13
	2.5 Conclusion	14
3	**Annuities and Perpetuities**	**17**
	3.1 Level Payments or Rate of Payment	17
	3.1.1 Annuities and Perpetuities Immediate	18
	3.1.2 Annuities and Perpetuities Due	20
	3.1.3 Several Payments per Year	22
	3.1.4 Continuous Annuities and Perpetuities	26
	3.1.5 Conclusion	27
	3.2 Arithmetic Growth	28
	3.2.1 Annuities and Perpetuities Immediate	28
	3.2.2 Annuities and Perpetuities Due	32
	3.2.3 Several Payments per Year	34
	3.2.4 Continuous Annuities and Perpetuities	39
	3.2.5 Conclusion	41
	3.3 Geometric Growth	41
	3.3.1 Annuities and Perpetuities Immediate	43
	3.3.2 Annuities and Perpetuities Due	45
	3.3.3 Special Cases	46
	3.3.4 Continuous Annuities and Perpetuities	47

	3.4	A Few Bonus Results	48
		3.4.1 Inverse Values	48
		3.4.2 Less Than a Payment per Year	49
		3.4.3 Asymptotic Results	50
		3.4.4 Deferred Annuities	51
		3.4.5 Rainbow Annuities	53

4 Loans — 56
- 4.1 Amortization Method 56
- 4.2 Sinking Fund Method 63

5 Bonds — 65
- 5.1 Definitions and Valuation 65
- 5.2 Zero-Coupon Bonds 69
- 5.3 Premium and Discount Bonds 70
  - 5.3.1 Main Definitions 71
  - 5.3.2 Book Value Evolution 72
  - 5.3.3 Comparison of Bonds and Loans 73
  - 5.3.4 Final Remark 76
- 5.4 Callable Bonds 76

6 General Cash Flows and Portfolios — 80
- 6.1 Rates of Return 80
- 6.2 Macaulay Duration and Convexity 83
- 6.3 Modified Duration and Convexity 85
- 6.4 Bonds 87
- 6.5 Level Annuities and Perpetuities 89
- 6.6 Increasing Annuities and Perpetuities 90
- 6.7 Stocks 91
- 6.8 Approximations 93
- 6.9 Forward Rates 95

7 Immunization — 98
- 7.1 Cash Flow Matching 99
- 7.2 Immunization Principles 100
- 7.3 A First Illustration 102
- 7.4 A Second Illustration 103
- 7.5 A Third Illustration 104

	7.6 A Further Generalization	105
8	**Interest Rate Swaps**	**107**
9	**Determinants of Interest Rates**	**116**
	9.1 Federal Reserve System	116
	9.2 Treasuries	118
	9.3 Interest Rates	119
Appendix		**124**

Chapter 1

Introduction

I wish I had had access to a book like this one when I took the exam FM of the Society of Actuaries (SOA). It contains all the formulas that you need to solve the official exercises of the SOA. I have tried to keep it very compact, theoretically solid, and not verbose.

Students who are not contemplating becoming actuaries but who are more interested in econometrics, finance, statistics, mathematics, or other fields, will also find this text useful.

I also recommend this book as a prerequisite to the students who are considering taking, or are in the process of taking, the CFA exams. Indeed, the fixed income and company valuation material studied in the CFA syllabus are based on the financial mathematics results provided in this book.

The order in which the contents of this book are presented mostly respects the order of the SOA syllabus. I made very few adjustments to this order and I did so only when pedagogical reasons supported the change.

This book does not just present the material; it helps you understand the foundations of the material. To keep the book compact, I purposely did not include exercises. On the contrary, this book should be used with the (long) series of exercises made freely available by the SOA.

The tables in the appendix link those exercises to the equations in the book. These tables can be a very convenient tool for providing hints for the exercises that you cannot solve – instead of going directly to the solutions.

I also suggest that you use this book in conjunction with the "BTDT (Been There Done That) Study Manual for Exam FM/2" written by Prof. Krzysztof Ostaszewski. This book can be obtained from http://smartURL.it/BTDT-FM. I also suggest that you take a look at http://smartURL.it/PassExamFM for a wealth of exercises with audio explanations.

This book is the second in a series that I am writing for actuarial associateship exams. In each of these books, I provide conceptual links between the contents of the various exams. This book was also written in such a way that you can use it throughout your career.

My best suggestion for those who are taking the exams is to avoid using flashcards. Why? Because if you need to quickly check a few days before an exam that you know your formulas, it just means you do not really know these formulas. More precisely, it means you have not mastered the concepts behind these formulas. So, if you are serious and if you have mastered your textbooks and have done hundreds of exercises, then flashcards are just completely useless because you know your formulas and the concepts behind them.

In the context of actuarial exams, I consider flashcards to be very dangerous for an additional reason: what you learn for an exam will be useful for the subsequent exams. Therefore, a deep mastery of the first exams is important when you start preparing for the most advanced exams. Again, this deep mastery is not attainable using "easy" tricks such as flashcards.

Financial Mathematics

At this time, I want to thank Li Shen, Xia Xu, and more specifically Xiaoshan Su, for their very kind help.

Before delving into this book, you should make sure that you are comfortable with classic calculus (summation, differentiation, integration, and so on). I will assume throughout the text that you know the following geometric series by heart. When the indexation starts from zero:

$$\sum_{k=0}^{n} q^k = \frac{1-q^{n+1}}{1-q} \quad \text{or} \quad \sum_{k=0}^{n-1} q^k = \frac{1-q^n}{1-q} \quad (1)$$

and

$$\sum_{k=0}^{+\infty} q^k = \frac{1}{1-q}, \quad (q < 1). \quad (2)$$

When the indexation starts from one:

$$\sum_{k=1}^{n} q^k = \frac{q-q^{n+1}}{1-q} \quad \text{or} \quad \sum_{k=1}^{n-1} q^k = \frac{q-q^n}{1-q} \quad (3)$$

and

$$\sum_{k=1}^{+\infty} q^k = \frac{q}{1-q}, \quad (q < 1). \quad (4)$$

A related result, which can be obtained by differentiating the left equality in Eq. (3), is

$$\sum_{k=1}^{n} k \, q^{k-1} = \frac{1-(n+1)\,q^n + n\,q^{n+1}}{(1-q)^2}. \quad (5)$$

The following result, which is not the value of a geometric series, is very important:

$$\sum_{k=1}^{n} k = \frac{n(n+1)}{2}. \quad (6)$$

I suggest that you read this book at least twice. You should **wait for your second reading to take a look at the footnotes and to do the suggested proofs.** Many proofs were left **intentionally** incomplete, with sufficient hints to be able to finish them by yourself. I cannot insist enough on how important it is that you do it. This will sharpen your understanding of financial mathematics and this will increase your computing speed, which is an important factor when you take exams.

To find out information about my publications, to get advice on the SOA and CAS exams, to kindly suggest corrections, or to subscribe to my monthly newsletter, see:

www.olivierlecourtois.com

Chapter 2

Time Value of Money

Money (usually) increases with time: if you deposit one hundred dollars at time 0 in a savings account, you will own (for instance) one hundred and two dollars in one year. The difference between the final and initial amounts in a savings account is called an **interest** or interest amount. When you express this capital gain in relative terms, you are dealing with an **interest rate**. Interest rates tell you how money can be moved forward and backward in time.

2.1 Interest Rates

Let S_0 be an initial amount of money and let S_t be a final amount of money. The quantity S_t is sometimes called an **amount function**, while $\frac{S_t}{S_0}$ is sometimes called an **accumulation function**. We are interested in examining how one can go from S_0 to S_t, or from S_t to S_0.

We define an **effective interest rate** i_t as the quantity associated with the period $(0, t)$ that satisfies [1]:

$$i_t = \frac{S_t - S_0}{S_0}. \qquad (7)$$

Specifically, when i is an *annual* effective interest rate, we

---
1. Other definitions exist.

have:
$$i = \frac{S_1 - S_0}{S_0}, \tag{8}$$
while a *semiannual* effective interest rate j can be defined as follows:
$$j = \frac{S_{\frac{1}{2}} - S_0}{S_0}, \tag{9}$$
and so on.

Therefore, an effective interest rate tells us how money increases in relative terms. However, it does not tell us about the most classic bank conventions for passing money forward or backward in time.

An important definition is that of a **simple interest rate** i_S, which allows you to build up value as follows:
$$S_t = S_0 \cdot (1 + i_S \cdot t), \tag{10}$$
where the **interest amount** I_t that is earned between times 0 and t is equal to
$$I_t = S_t - S_0 = S_0 \cdot i_S \cdot t, \tag{11}$$
and an effective interest rate can be computed as follows:
$$i_t = \frac{S_t - S_0}{S_0} = \frac{I_t}{S_0} = i_S \cdot t. \tag{12}$$

You can easily check that $S_{2t} - S_t = S_0 \cdot i_S \cdot t = S_t - S_0$, so that the interest amount that is received between two consecutive periods of equal length is equal. This means that with simple interest rates, you do not receive interest on previously earned interest. This is not what happens in the real world, where you receive interest on previously earned interest.

From now on (and until the end of this book) we consider **compound, also called compounded, interest rates**, that is, we consider situations where interest on interest is computed. Therefore, we consider situations where $S_{2t} - S_t \neq S_t - S_0$.

Specifically, being given an **annual compound interest rate** i, we **compound** money as follows:
$$S_n = S_0 \cdot (1 + i)^n, \tag{13}$$

Financial Mathematics

where n is a number of years.

Assume for instance that you invest $S_0 = 100$ at time 0 and that the annual compound interest rate is $i = 5\%$. If you wait for $n = 3$ years, you own at time $t = 3$ an amount that is equal to $S_3 = 100 \cdot (1.05)^3 = 115.7625$. The interest is $S_3 - S_0 = 15.7625$, which is bigger than $15 = 3 \cdot 5\% \cdot 100$. Interest on interest has been earned.

It turns out that the annual compound interest rate i used in Eq. (13) is an annual effective interest rate. Indeed,

$$\frac{S_1 - S_0}{S_0} = \frac{S_0 \cdot (1+i) - S_0}{S_0} = i. \tag{14}$$

From now on, we will use the terms "compound rate" and "effective rate" interchangeably. But you have observed that the latter term is broader because you can also compute an effective interest rate when you are given a simple interest rate.

Moving money backward in time, also called **discounting**, is performed by inverting Eq. (13):

$$S_0 = \frac{S_n}{(1+i)^n}. \tag{15}$$

Effective (compound) interest rates are not always annual. Let i and j be any two effective interest rates for the same financial product, where we only know that i covers m times the period covered by j. Then, we have the following important relation used in many exercises:

$$1 + i = (1+j)^m, \tag{16}$$

where j is also sometimes denoted as $j^{(m)}$.

For instance, let i be an annual effective interest rate and j, or $j^{(2)}$, be a semiannual effective interest rate. In this simple case, we have:

$$1 + i = (1+j)^2. \tag{17}$$

Another useful interest rate is the **effective discount rate**, or **rate of discount** [2], which is the effective interest rate discounted by itself:

$$d = \frac{i}{1+i}. \tag{20}$$

The reciprocal formula is also important to know and is expressed as follows:

$$i = \frac{d}{1-d}. \tag{21}$$

Discount rates allow us to discount money in the following way:

$$S_0 = S_n \cdot (1-d)^n, \tag{22}$$

while compounding is performed as follows:

$$S_n = \frac{S_0}{(1-d)^n}, \tag{23}$$

where you should take time to compare these two equations with Eqs (13) and (15).

Discount rates are useful for pricing products whose interest is paid *ex ante*. For example, an amount S_1 that is reimbursed at time 1 is associated with a smaller amount $S_1(1-d)$ that is borrowed at time 0.

A key concept in financial mathematics is that of a **discount factor** v, which is defined by

$$v = \frac{1}{1+i}, \tag{24}$$

where the denominator is sometimes called an **interest factor**.

---

2. When we consider a reference period $(0, t)$ that is not necessarily annual, the effective discount rate d_t is defined in full generality as follows:

$$d_t = \frac{S_t - S_0}{S_t}. \tag{18}$$

When the reference period is annual, we have:

$$d = \frac{S_1 - S_0}{S_1}, \tag{19}$$

which allows us to recover Eq. (20), using Eq. (8). Please check it.

This quantity allows you to directly discount a sum of money over a unitary time period, for instance from time 2 to time 1 (because $S_1 = v\, S_2$), from time 1 to time 0 (because $S_0 = v\, S_1$), and so on. Discounting over several periods is achieved as follows:

$$S_0 = S_n \cdot v^n. \tag{25}$$

Eq. (24) can be reversed to express i as a function of v:

$$i = \frac{1-v}{v}. \tag{26}$$

We also have a simple link between v and d:

$$d = 1 - v. \tag{27}$$

Combining Eqs (26) and (27), we obtain:

$$d = i\, v, \tag{28}$$

which is an alternative form of Eq. (20).

2.2 How to Deal with Sub-Periods

Next, we define a (most often annual) **nominal interest rate** $i^{(m)}$ **convertible, or compounded, m^{thly}, or m times per period**. This interest rate allows us to go from S_0 to S_1 as follows:

$$S_1 = S_0 \cdot \left(1 + \frac{i^{(m)}}{m}\right)^m, \tag{29}$$

where we compound money m consecutive times over subperiods of length $\frac{1}{m}$.

From this equation and from Eq. (13) we deduce that

$$1 + i = \left(1 + \frac{i^{(m)}}{m}\right)^m. \tag{30}$$

2.3 Continuous Compounding

Another way to compound money is with the annual **force of interest** δ:
$$S_1 = S_0 \cdot e^\delta, \tag{45}$$
where, more generally, we have:
$$S_t = S_0 \cdot e^{\delta t}, \tag{46}$$
and where t can take any positive real value.

This is also called **continuous compounding**, where at each point of time an infinitesimal interest is received on the infinitesimal interest earned immediately before. Discounting with a force of interest is achieved as follows:
$$S_0 = S_t \cdot e^{-\delta t}. \tag{47}$$

This way of compounding and discounting money is classic in derivatives pricing (an important part of exam IFM).

It is important to know the link between the rates δ and i:
$$1 + i = e^\delta \quad \Leftrightarrow \quad \delta = \ln(1 + i), \tag{48}$$
and the link between δ and v:
$$v = e^{-\delta} \quad \Leftrightarrow \quad \delta = -\ln(v). \tag{49}$$

For a non-constant force of interest δ_s, it is possible to generalize Eq. (46) as follows:
$$S_t = S_0 \cdot e^{\int_0^t \delta_s \, ds}, \tag{50}$$
where at each future point of time s, a different interest rate δ_s is used.

Note that it is possible[3] to reverse this formula by computing $\frac{S_t'}{S_t}$. We obtain an expression for the force of interest δ_t:
$$\delta_t = \frac{S_t'}{S_t}, \tag{51}$$

---

3. Do it as an exercise, using the well-known form of the derivative of the exponential of a function, and the fact that the derivative of the primitive of a function is simply this function.

which expresses the force of interest as the ratio of the derivative of the compound amount to the compound amount itself.

Of course, we also have:
$$\delta_t = \frac{\left(\frac{S_t}{S_0}\right)'}{\frac{S_t}{S_0}}, \tag{52}$$

which states that the force of interest is the ratio of the derivative of the accumulation function to the accumulation function itself.

All of the examples above can be rewritten with the **accumulation function** f, which satisfies:
$$S_t = S_0 \cdot f(t). \tag{53}$$

We can relate S_t and S_u as follows:
$$S_u = S_t \cdot \frac{f(u)}{f(t)}, \tag{54}$$

so that for instance:
$$S_u = S_t \cdot e^{\int_t^u \delta_s ds}. \tag{55}$$

The interest received between general times $t < u$ is
$$S_u - S_t = S_t \cdot \left(\frac{f(u)}{f(t)} - 1\right), \tag{56}$$

so that, between times 0 and T,
$$S_T - S_0 = S_0 \cdot (f(T) - 1) = S_0 \cdot \left(e^{\int_0^T \delta_s ds} - 1\right). \tag{57}$$

2.4 Pricing

The **present value** V_t at time t of a continuous stream of payments p_u, for $u \in [t, T]$, is
$$V_t = \int_t^T p_u \, e^{-\int_t^u \delta_s ds} \, du, \tag{58}$$

where T may be infinite.

The **future value** V_T at time T of a continuous stream of payments p_u, for $u \in [t, T]$, is

$$V_T = \int_t^T p_u \, e^{\int_u^T \delta_s ds} \, du. \tag{59}$$

A key conceptual tool in the corporate finance and investments industries is the **net present value** NPV(0) at time 0 of a series of n cash flows CF_{t_j} paid at times t_j. This tool is defined as follows:

$$\text{NPV}(0) = \sum_{j=1}^n \text{CF}_{t_j} \, v^{t_j}, \tag{60}$$

and allows an analyst to evaluate a project, a company, an asset, a security, ...

By analogy, the **net future value** NFV(T) at time T of a series of n cash flows CF_{t_j} paid at times t_j is equal to

$$\text{NFV}(T) = \sum_{j=1}^n \text{CF}_{t_j} \, (1+i)^{T-t_j}. \tag{61}$$

Finally, an **Equation of value** is simply a way of equating in the present or in the future a value that is computed, discounted, or compounded, using two different methods. It can also amount to computing an NPV or an NFV, where the cash flows can be positive and negative, and to equating this NPV or NFV to zero.

2.5 Conclusion

The force of interest and the other main interest and discount rates are ordered in a strict way:

$$d < d^{(p)} < d^{(q)} < \delta < i^{(q)} < i^{(p)} < i, \tag{62}$$

where p and q are two integers that satisfy $p < q$, and where $\lim_{p \to +\infty} d^{(p)} = \delta$ and $\lim_{p \to +\infty} i^{(p)} = \delta$.

Assume for instance that $\delta = 5\%$. Then, we can compute $i = 5.13\%$, $d = 4.88\%$, $i^{(2)} = 5.06\%$, $d^{(2)} = 4.94\%$, $i^{(12)} = 5.01\%$, and $d^{(12)} = 4.99\%$.

See Table 1 for a summary of the relations between the main rates.

	i	v	d	δ	$i^{(m)}$	$d^{(m)}$
i		$\frac{1}{1+i}$	$\frac{i}{1+i}$	$\ln(1+i)$	$m\left((1+i)^{\frac{1}{m}} - 1\right)$	$m\left(1 - (1+i)^{-\frac{1}{m}}\right)$
v	$\frac{1}{v} - 1$		$1 - v$	$-\ln(v)$	$m\left(v^{-\frac{1}{m}} - 1\right)$	$m\left(1 - v^{\frac{1}{m}}\right)$
d	$\frac{d}{1-d}$	$1 - d$		$-\ln(1-d)$	$m\left((1-d)^{-\frac{1}{m}} - 1\right)$	$m\left(1 - (1-d)^{\frac{1}{m}}\right)$
δ	$e^{\delta} - 1$	$e^{-\delta}$	$1 - e^{-\delta}$		$m\left(e^{\frac{\delta}{m}} - 1\right)$	$m\left(1 - e^{-\frac{\delta}{m}}\right)$
$i^{(m)}$	$\left(1 + \frac{i^{(m)}}{m}\right)^m - 1$	$\left(1 + \frac{i^{(m)}}{m}\right)^{-m}$	$1 - \left(1 + \frac{i^{(m)}}{m}\right)^{-m}$	$m \ln\left(1 + \frac{i^{(m)}}{m}\right)$		$\frac{i^{(m)}}{1 + \frac{i^{(m)}}{m}}$
$d^{(m)}$	$\left(1 - \frac{d^{(m)}}{m}\right)^{-m} - 1$	$\left(1 - \frac{d^{(m)}}{m}\right)^m$	$1 - \left(1 - \frac{d^{(m)}}{m}\right)^m$	$-m \ln\left(1 - \frac{d^{(m)}}{m}\right)$	$\frac{d^{(m)}}{1 - \frac{d^{(m)}}{m}}$	

Table 1 – Useful Interest Rate Relations.

Chapter 3

Annuities and Perpetuities

This chapter deals with annuities and perpetuities **certain**. These products pay series of cash flows over a **predetermined time span**, which is finite in the case of annuities and infinite in the case of perpetuities. The payments occur for sure, whether the recipient is alive or dead: should he or she die, the payments are made to his or her heirs. We do not consider **life** annuities and perpetuities, whose payments stop when the recipient dies. You will study these financial products when you prepare for the exam LTAM of the SOA. Reaching a good mastery of annuities and perpetuities certain will help you a lot later on when you study life annuities and perpetuities.

3.1 Level Payments or Rate of Payment

We start by considering the situation where payments are **level**, meaning **constant**. We also consider in this section the case where a level rate of payment is offered by the annuity or perpetuity.

3.1.1 Annuities and Perpetuities Immediate

The **term** of an annuity or a perpetuity is the total time span of this financial product. The term of annuities is finite, while that of perpetuities is infinite.

We first consider the case of annuities and perpetuities **immediate**. These financial products make payments **in arrears** at the **end** of (usually) equal-length periods [1].

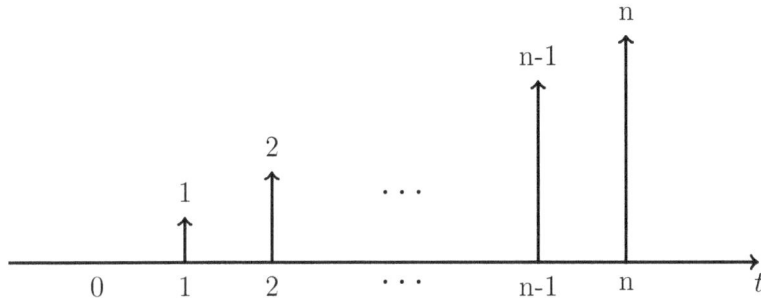

Figure 1 – Cash Flow Timeline of an Annuity Immediate

Let us consider an annuity that makes n payments of 1 at times $1, \cdots, n$. See Fig. 1 for a cash flow timeline. The **present value** $a_{\overline{n}|i}$ **of this annuity immediate** is computed by discounting and summing the n unitary payments:

$$a_{\overline{n}|i} = \sum_{k=1}^{n} 1 \cdot v^k. \qquad (63)$$

Therefore,

$$a_{\overline{n}|i} = \frac{1-v^n}{i}, \qquad (64)$$

as you should check using Eqs (3) and (26). Pay attention to the notation $a_{\overline{n}|i}$. You will use it a lot and much more sophisticated notation of the same type will be introduced for exam LTAM.

---
1. You should remember that the name of an annuity immediate is counterintuitive: this financial product does NOT make immediate payments, because the first payment occurs at the end of the first period and not at time 0.

Now, the **future value $s_{\overline{n}|i}$ of an annuity immediate** is computed at time n as follows:

$$s_{\overline{n}|i} = \sum_{k=0}^{n-1} 1 \cdot (1+i)^k, \qquad (65)$$

by compounding the n unitary payments made at times $1, \cdots, n$ to time n. In this sum, $k = 0$ corresponds to the last payment made at time n. There is no need to compute its value at time n because it is simply worth itself: $1 \cdot (1+i)^0 = 1$. Then, $k = 1$ corresponds to the penultimate payment that is made at time $n-1$ and that should be compounded once. And so on for larger values of k.

A closed-form formula for this future value exists:

$$s_{\overline{n}|i} = \frac{(1+i)^n - 1}{i} = \frac{v^{-n} - 1}{i}, \qquad (66)$$

as you can check using Eqs (65), (1), and (24). It is also possible to directly go from Eq. (64) to Eq. (66) by a multiplication with the compounding factor $(1+i)^n$.

Indeed, it is important to observe that one can directly go from the present value of an annuity immediate to its future value by compounding the value of the product from time 0 to time n:

$$s_{\overline{n}|i} = (1+i)^n \, a_{\overline{n}|i}. \qquad (67)$$

Perpetuities are just like annuities except that they last forever. Therefore, the **present value $a_{\overline{\infty}|i}$ of a perpetuity immediate** is computed as

$$a_{\overline{\infty}|i} = \sum_{k=1}^{\infty} v^k = \frac{v}{1-v} = \frac{1}{i}, \qquad (68)$$

where the second equality is a consequence of Eq. (4) and the third equality follows from Eq. (26).

3.1.2 Annuities and Perpetuities Due

We now consider the case of annuities and perpetuities **due**. These products make payments **in advance** at the **beginning** of (usually) equal-length periods. See Fig. 2 for a cash flow timeline. The **present value** $\ddot{a}_{\overline{n}|i}$ **of an annuity due** that has a term of n years is

$$\ddot{a}_{\overline{n}|i} = \sum_{k=0}^{n-1} v^k, \tag{69}$$

where the first payment occurs at time 0 and the last payment occurs at time $n-1$.

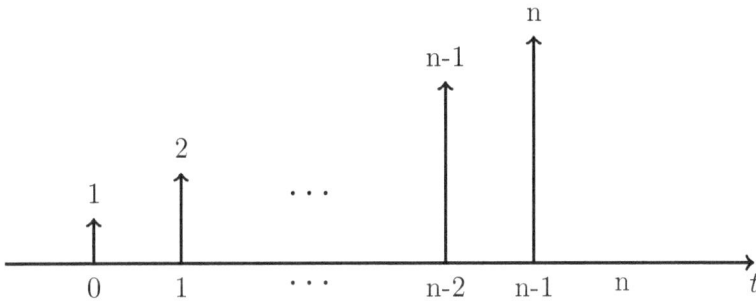

Figure 2 – Cash Flow Timeline of an Annuity Due

The previous formula has a simple closed-form expression:

$$\ddot{a}_{\overline{n}|i} = \frac{1-v^n}{d}, \tag{70}$$

as you can check using Eqs (69), (1), and (27).

You should remember that the present values of annuities immediate and annuities due are related as follows[2]:

$$\ddot{a}_{\overline{n}|i} = (1+i)\, a_{\overline{n}|i}, \tag{71}$$

where the present value of an annuity due is larger than the present value of an otherwise equivalent annuity immediate.

---

2. This result can be checked by dividing Eq. (70) by Eq. (64) and by using Eq. (20).

Indeed, because the payments of annuities due occur one period earlier than the payments of annuities immediate, they should be discounted one less period. Therefore, they should be bigger by a factor of $(1+i)$.

Another important result is as follows:

$$\ddot{a}_{\overline{n}|i} = 1 + a_{\overline{n-1}|i}, \qquad (72)$$

where we simply write that when we separate out the first payment of an annuity due, we are left with an annuity immediate with one less payment.

By combining Eqs (71) and (72), we are able to construct the following result:

$$a_{\overline{n}|i} = v \left(1 + a_{\overline{n-1}|i}\right), \qquad (73)$$

which can be used to compute the values of annuities immediate by induction.

Let us now come to the computation of the **future value** $\ddot{s}_{\overline{n}|i}$ **of an annuity due**. We write:

$$\ddot{s}_{\overline{n}|i} = \sum_{k=1}^{n} (1+i)^k, \qquad (74)$$

where, as for the computation of the future value of an annuity immediate, the first values of k correspond to the last payments of the annuity. The future value of an annuity due can also be written as follows:

$$\ddot{s}_{\overline{n}|i} = \frac{(1+i)^n - 1}{d} = \frac{v^{-n} - 1}{d}, \qquad (75)$$

as you can check using Eqs (74), (3), (20), and (24). It is also possible to directly go from Eq. (70) to Eq. (75) by a multiplication with the compounding factor $(1+i)^n$.

Indeed, as for annuities immediate, we can easily compound the values of annuities due from time 0 to time n:

$$\ddot{s}_{\overline{n}|i} = (1+i)^n \, \ddot{a}_{\overline{n}|i}. \qquad (76)$$

Using Eqs (67) and (76) and compounding Eq. (71) with $(1+i)^n$ allows us to write:

$$\ddot{s}_{\overline{n}|i} = (1+i)\, s_{\overline{n}|i}, \tag{77}$$

which says that the future value of an annuity due is larger than the future value of an otherwise equivalent annuity immediate.

Further, by compounding Eq. (72) by $(1+i)^n$, we obtain:

$$\ddot{s}_{\overline{n}|i} = (1+i)^n + (1+i)\cdot s_{\overline{n-1}|i}, \tag{78}$$

where you should be careful that $\ddot{s}_{\overline{n}|i}$ is a value at time n, while $s_{\overline{n-1}|i}$ is a value at time $n-1$. In order to compare these two values, the latter should be compounded once to be taken to time n, as is performed in Eq. (78). Then, clearly, $(1+i)^n$ is the future value at time n of an isolated payment of 1 made at time 0.

Finally, observe that the **present value $\ddot{a}_{\overline{\infty}|i}$ of a perpetuity due** is equal to

$$\ddot{a}_{\overline{\infty}|i} = \sum_{k=0}^{\infty} v^k = \frac{1}{1-v} = \frac{1}{d}, \tag{79}$$

as you can check using Eqs (2) and (27). Note that this formula can also be obtained by taking the limit $n \to +\infty$ in Eq. (70) and by using the fact that $\lim_{n \to +\infty} v^n = 0$.

3.1.3 Several Payments per Year

We first study an annuity immediate that pays *one currency unit* m-thly, so m times per year, over n years. In total, this financial product makes $n \cdot m$ payments. The time lag between any two payments is $\frac{1}{m}$. The kth cash flow of this annuity occurs at time $\frac{k}{m}$. See Fig. 3 for a cash flow timeline.

Let $j^{(m)} = (1+i)^{\frac{1}{m}} - 1$ be the effective interest rate that covers a period of length $\frac{1}{m}$ and let $v_m = \frac{1}{1+j^{(m)}} = v^{\frac{1}{m}}$ be the discount factor over the same period. By analogy with Eq. (63), we compute the present value of this annuity as follows:

$$a_{\overline{n\cdot m}|j^{(m)}} = \sum_{k=1}^{nm} v^{\frac{k}{m}} = \sum_{k=1}^{nm} (v_m)^k = \frac{1-(v_m)^{nm}}{j^{(m)}}, \tag{80}$$

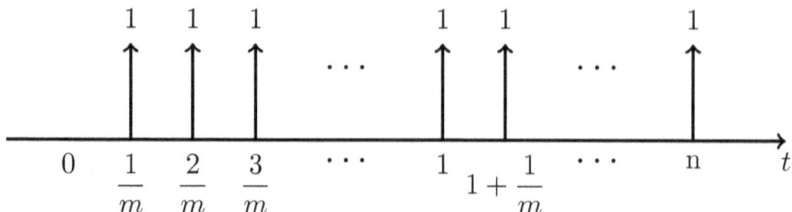

Figure 3 – Cash Flow Timeline Associated with $a_{\overline{n \cdot m}|j^{(m)}}$

where in the first sum we simply discount the nm cash flows from their occurrence time $\frac{k}{m}$ to time 0. The second sum is a direct consequence of $v_m = v^{\frac{1}{m}}$. Finally, the rightmost expression is obtained by analogy with Eq. (64).

Recall from Eq. (34) that $j^{(m)} = \frac{i^{(m)}}{m}$. From this fact and from Eq. (80), we deduce that the **present value of an annuity immediate that makes a payment of 1 (!) m times per year for n years** is equal to

$$a_{\overline{n \cdot m}|j^{(m)}} = a_{\overline{n \cdot m}|\frac{i^{(m)}}{m}} = \frac{1 - v^n}{\frac{i^{(m)}}{m}}. \tag{81}$$

The **present value of an annuity immediate that makes a payment of $\frac{1}{m}$ (!) m times per year for n years** (so that, again, nm payments are made ; only the size of the cash flows has changed), is denoted by $a_{\overline{n}|i}^{(m)}$. This annuity pays $1 = m \cdot \frac{1}{m}$ currency unit per year. See Fig. 4 for a cash flow timeline. The present value of this annuity can be readily computed as follows:

$$a_{\overline{n}|i}^{(m)} = \frac{1}{m} a_{\overline{n \cdot m}|j^{(m)}} = \frac{1 - v^n}{i^{(m)}}. \tag{82}$$

The future value at time n of this annuity immediate is computed by compounding its present value with $(1+i)^n$. We obtain:

$$s_{\overline{n}|i}^{(m)} = \frac{(1+i)^n - 1}{i^{(m)}} = \frac{v^{-n} - 1}{i^{(m)}}. \tag{83}$$

Finally, a perpetuity immediate that makes m payments of $\frac{1}{m}$ per year is priced by taking the limit $n \to +\infty$ in the pricing

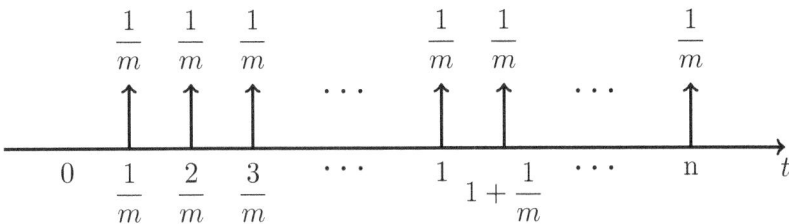

Figure 4 – Cash Flow Timeline Associated with $a_{\overline{n}|i}^{(m)}$

equation (82). We obtain:

$$a_{\overline{\infty}|i}^{(m)} = \frac{1}{i^{(m)}}. \qquad (84)$$

Let us now come to the study of financial products whose first payment occurs at time 0. By analogy with Eq. (70), we compute the **present value of an annuity due that pays one currency unit m times per year over n years** as follows:

$$\ddot{a}_{\overline{n\cdot m}|j^{(m)}} = \sum_{k=0}^{nm-1} v_m^{\frac{k}{m}} = \frac{1 - (v_m)^{n \cdot m}}{d_m} = \frac{1 - \left(\frac{1}{1+j^{(m)}}\right)^{n \cdot m}}{\frac{j^{(m)}}{1+j^{(m)}}}, \qquad (85)$$

where nm payments are made at times $\frac{k}{m}$, with $k = 0, \ldots, nm-1$, and $d_m = \frac{j^{(m)}}{1+j^{(m)}} = 1 - v_m$ is the effective discount rate over a period of length $\frac{1}{m}$. See Fig. 5 for a cash flow timeline.

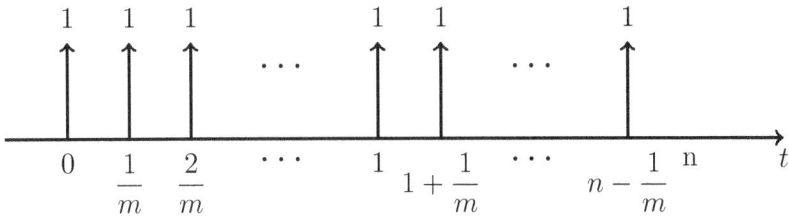

Figure 5 – Cash Flow Timeline Associated with $\ddot{a}_{\overline{n\cdot m}|j^{(m)}}$

Because $d_m = \frac{d^{(m)}}{m}$, we can rewrite Eq. (85) as follows:

$$\ddot{a}_{\overline{n \cdot m}| j^{(m)}} = \ddot{a}_{\overline{n \cdot m}| \frac{i^{(m)}}{m}} = \frac{1 - v^n}{\frac{d^{(m)}}{m}}. \tag{86}$$

Let us now price an annuity that makes a **payment of $\frac{1}{m}$ at times** $0, \frac{1}{m}, \frac{2}{m}, \frac{3}{m}, \ldots, \frac{nm-1}{m} = n - \frac{1}{m}$. In total, nm payments are made and this annuity is said to pay $1 = m \cdot \frac{1}{m}$ currency unit per year. The present value of this annuity is denoted by $\ddot{a}_{\overline{n}|i}^{(m)}$. See its cash flow timeline in Fig. 6. It can be priced as follows:

$$\ddot{a}_{\overline{n}|i}^{(m)} = \frac{1}{m} \ddot{a}_{\overline{n \cdot m}| j^{(m)}} = \frac{1 - v^n}{d^{(m)}}. \tag{87}$$

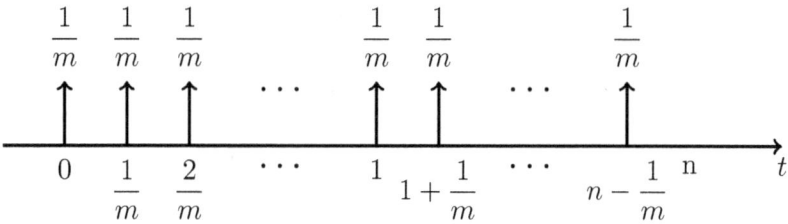

Figure 6 – Cash Flow Timeline Associated with $\ddot{a}_{\overline{n}|i}^{(m)}$

As before, we readily derive the future value of this financial product by multiplying its present value by $(1+i)^n$. We obtain:

$$\ddot{s}_{\overline{n}|i}^{(m)} = \frac{(1+i)^n - 1}{d^{(m)}} = \frac{v^{-n} - 1}{d^{(m)}}. \tag{88}$$

A perpetuity due with m payments of $\frac{1}{m}$ per year is priced by taking the limit $n \to +\infty$ in Eq. (87). We obtain:

$$\ddot{a}_{\overline{\infty}|i}^{(m)} = \frac{1}{d^{(m)}}. \tag{89}$$

By comparing Eqs (82) and (87), we can relate the present values of annuities due and immediate that make several payments per year:

$$\ddot{a}_{\overline{n}|i}^{(m)} = \left(1 + \frac{i^{(m)}}{m}\right) a_{\overline{n}|i}^{(m)}, \tag{90}$$

where a similar link exists between the future values of these financial products, and also between the present values of perpetuities due and immediate that make several payments per year.

Finally, note that when you learn the material of this section, you also get prepared for exam LTAM, where life annuities that make several payments per period are used extensively.

3.1.4 Continuous Annuities and Perpetuities

Continuous annuities and perpetuities are hypothetical objects that provide a continuous stream of cash. Obviously, it is not reasonable to assume that a counterparty can pay to another one an infinitesimally small amount of cash at every point of time during the life of a financial product. However, nothing prevents us from pricing such hypothetical products. By doing so, we obtain very useful approximate formulas for situations where the frequency of payments is high.

We start by computing the **present value of a continuous annuity** that pays a rate of cash normalized to 1 between times 0 and T. We obtain:

$$\bar{a}_{\overline{T}|i} = \int_0^T 1 \cdot v^t \, dt = \frac{1 - v^T}{\delta} = \frac{1 - e^{-\delta T}}{\delta}, \qquad (91)$$

where the rate of cash of 1 that is paid at time t is discounted from time t to time 0 using v^t. Because t can take any real value in $(0, T)$, the sum in Eq. (63) is replaced with an integral.

The **future value of a continuous annuity** can be derived by compounding with $(1+i)^T$ its present value from time 0 to time T:

$$\bar{s}_{\overline{T}|i} = (1+i)^T \, \bar{a}_{\overline{T}|i} = \frac{v^{-T} - 1}{\delta} = \frac{e^{\delta T} - 1}{\delta}. \qquad (92)$$

Taking the limit $T \to +\infty$ in Eq. (91), we can also derive the **present value of a continuous perpetuity**:

$$\bar{a}_{\overline{\infty}|i} = \int_0^{+\infty} v^s \, ds = \frac{1}{\delta}. \qquad (93)$$

Financial Mathematics

3.1.5 Conclusion

We show in Table 2 a summary of the present and future values of annuities and of the present values of perpetuities.

$a_{\overline{n}|i}$	$a_{\overline{n}|i}^{(m)}$	$\bar{a}_{\overline{T}|i}$	$\ddot{a}_{\overline{n}|i}^{(m)}$	$\ddot{a}_{\overline{n}|i}$
$\frac{1-v^n}{i}$	$\frac{1-v^n}{i^{(m)}}$	$\frac{1-v^T}{\delta}$	$\frac{1-v^n}{d^{(m)}}$	$\frac{1-v^n}{d}$
12.329	12.616	12.642	12.669	12.961
$s_{\overline{n}|i}$	$s_{\overline{n}|i}^{(m)}$	$\bar{s}_{\overline{T}|i}$	$\ddot{s}_{\overline{n}|i}^{(m)}$	$\ddot{s}_{\overline{n}|i}$
$\frac{v^{-n}-1}{i}$	$\frac{v^{-n}-1}{i^{(m)}}$	$\frac{v^{-T}-1}{\delta}$	$\frac{v^{-n}-1}{d^{(m)}}$	$\frac{v^{-n}-1}{d}$
33.514	34.294	34.366	34.437	35.232
$a_{\overline{\infty}|i}$	$a_{\overline{\infty}|i}^{(m)}$	$\bar{a}_{\overline{\infty}|i}$	$\ddot{a}_{\overline{\infty}|i}^{(m)}$	$\ddot{a}_{\overline{\infty}|i}$
$\frac{1}{i}$	$\frac{1}{i^{(m)}}$	$\frac{1}{\delta}$	$\frac{1}{d^{(m)}}$	$\frac{1}{d}$
19.504	19.958	20.000	20.042	20.504

Table 2 – Valuation of Annuities and Perpetuities. The Illustrations are Computed Using $T = n = 20$, $m = 12$, and $\delta = 5\%$.

Very useful and simple ratios can be derived by comparing any two columns from Table 2. Consider for instance the second and fifth column of the table. We readily obtain:

$$\frac{a_{\overline{n}|i}^{(m)}}{\ddot{a}_{\overline{n}|i}} = \frac{s_{\overline{n}|i}^{(m)}}{\ddot{s}_{\overline{n}|i}} = \frac{a_{\overline{\infty}|i}^{(m)}}{\ddot{a}_{\overline{\infty}|i}} = \frac{d}{i^{(m)}}, \qquad (94)$$

where you should be able to derive similar ratios using other columns of the table. These ratios are useful when you are given three quantities (for example $a_{\overline{n}|i}^{(m)}$, d, and $i^{(m)}$) and you want to compute a fourth related quantity (in this example $\ddot{a}_{\overline{n}|i}$).

Interest rate rankings allow us to rank annuity present values. Using the sequence of inequalities $d < d^{(m)} < \delta < i^{(m)} < i$, we obtain [3]:

$$a_{\overline{n}|i} < a_{\overline{n}|i}^{(m)} < \bar{a}_{\overline{n}|i} < \ddot{a}_{\overline{n}|i}^{(m)} < \ddot{a}_{\overline{n}|i}. \qquad (95)$$

---

3. This can be easily checked using the formulas from Table 2.

A similar ranking holds for annuity future values:

$$s_{\overline{n}|i} < s_{\overline{n}|i}^{(m)} < \bar{s}_{\overline{n}|i} < \ddot{s}_{\overline{n}|i}^{(m)} < \ddot{s}_{\overline{n}|i}, \qquad (96)$$

and we also have for perpetuity present values:

$$a_{\overline{\infty}|i} < a_{\overline{\infty}|i}^{(m)} < \bar{a}_{\overline{\infty}|i} < \ddot{a}_{\overline{\infty}|i}^{(m)} < \ddot{a}_{\overline{\infty}|i}. \qquad (97)$$

3.2 Arithmetic Growth

We now consider a class of products whose payments increase or decrease by a constant amount: we say that we are in a situation of **arithmetic growth or decline**.

3.2.1 Annuities and Perpetuities Immediate

We first compute the present value of an annuity immediate that starts distributing K and whose next $n-1$ payments sequentially increase by L. For illustration, the first payment of this annuity is K at time 1, its second payment is $K+L$ at time 2, its third payment is $K+2L$ at time 3, and so on until its last payment that is worth $K+(n-1)L$ at time n.

Owning this financial product is equivalent to receiving n payments of K between times 1 and n, plus $n-1$ payments of L between times 2 and n, plus $n-2$ payments of L between times 3 and n, and so on. Therefore, the present value $(Ia)_{\overline{n}|i}^{[K,L]}$ of this arithmetically increasing annuity immediate can be computed as follows:

$$(Ia)_{\overline{n}|i}^{[K,L]} = K\, a_{\overline{n}|i} + L \sum_{k=1}^{n-1} v^k\, a_{\overline{n-k}|i}, \qquad (98)$$

where $k=1$ in the sum corresponds to the $n-1$ payments of L that are made between times 2 and n, that are worth $L\, a_{\overline{n-1}|i}$ at time 1, and that need to be discounted once with v to be valued at time 0, and so on.

We can show that the **present value of this arithmetically increasing annuity immediate** is equal to [4]

$$(Ia)_{\overline{n}|i}^{[K,L]} = K\, a_{\overline{n}|i} + L\, \frac{a_{\overline{n}|i} - n\, v^n}{i}. \tag{101}$$

Multiplying by $(1+i)^n$, we can derive the future value of this financial product, which is equal to

$$(Is)_{\overline{n}|i}^{[K,L]} = K\, s_{\overline{n}|i} + L\, \frac{s_{\overline{n}|i} - n}{i}, \tag{102}$$

while the present value of this arithmetically increasing annuity immediate, derived by taking the limit $n \to +\infty$ in Eq. (101), is equal to

$$(Ia)_{\overline{\infty}|i}^{[K,L]} = K\, a_{\overline{\infty}|i} + L\, \frac{a_{\overline{\infty}|i}}{i} = \frac{K}{i} + \frac{L}{i^2}, \tag{103}$$

where we use the fact that $\lim_{n \to +\infty} n\, v^n = 0$.

When $K = L = 1$, we are in the situation of a **standard arithmetically increasing annuity immediate**. See Fig. 7 for a cash flow timeline. The present value $(Ia)_{\overline{n}|i}^{[1,1]}$ of this financial product, which we more simply denote by $(Ia)_{\overline{n}|i}$, is equal to [5]

$$(Ia)_{\overline{n}|i} = \sum_{k=1}^{n} k\, v^k = \frac{\ddot{a}_{\overline{n}|i} - n\, v^n}{i}, \tag{104}$$

from which we deduce the following future value:

$$(Is)_{\overline{n}|i} = \frac{\ddot{s}_{\overline{n}|i} - n}{i}. \tag{105}$$

---

4. Based on Eqs (98) and (64), we compute:

$$(Ia)_{\overline{n}|i}^{[K,L]} = K\, a_{\overline{n}|i} + L \sum_{k=1}^{n-1} v^k\, \frac{1 - v^{n-k}}{i} = K\, a_{\overline{n}|i} + L \sum_{k=1}^{n-1} \frac{v^k - v^n}{i}. \tag{99}$$

Then, we insert a zero into the sum and we write:

$$(Ia)_{\overline{n}|i}^{[K,L]} = K\, a_{\overline{n}|i} + L \sum_{k=1}^{n} \frac{v^k - v^n}{i}. \tag{100}$$

Finally, using Eq. (63), we obtain Eq. (101).

5. You can use Eqs (71) and (101) to compute it.

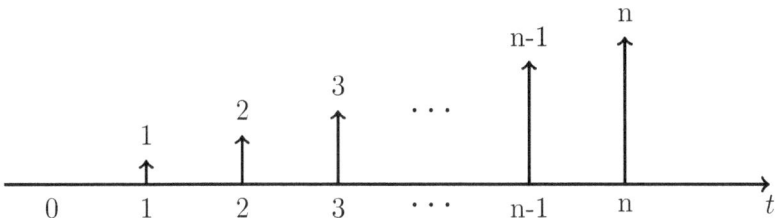

Figure 7 – Cash Flow Timeline of a Standard Arithmetically Increasing Annuity Immediate

Be careful of the double dots in Eqs (104) and (105): the arithmetically increasing annuity we evaluate is *immediate*, but its present and future values depend on the present and future values of an annuity *due*.

Using Eqs (104) and (5), we are able to derive an alternative expression for the present value of a standard arithmetically increasing annuity immediate:

$$(Ia)_{\overline{n}|i} = \frac{v}{(1-v)^2}\left[1 - (n+1)\,v^n + n\,v^{n+1}\right]. \tag{106}$$

Then, for a **standard arithmetically increasing perpetuity immediate**, we compute

$$(Ia)_{\overline{\infty}|i} = \frac{\ddot{a}_{\overline{\infty}|i}}{i} = \frac{1}{i\,d} = \frac{1+i}{i^2}, \tag{107}$$

where we use the fact that $\lim\limits_{n\to+\infty} nv^n = 0$ in Eq. (104).

Finally, let us consider the case of a **standard arithmetically decreasing annuity**. Such a product makes a first payment of n at time 1, while each of its $n-1$ subsequent payments is smaller than the previous one by one currency unit. Therefore, the last payment is equal to 1 at time n. See Fig. 8 for a cash flow timeline.

This arithmetically decreasing annuity is valued by setting $K = n$ and $L = -1$ in Eq. (101), from which we deduce the following initial value:

$$(Da)_{\overline{n}|i} = \frac{n - a_{\overline{n}|i}}{i}, \tag{108}$$

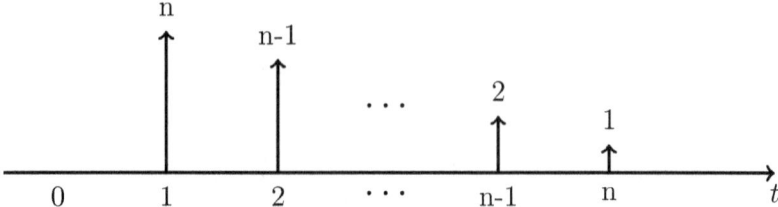

Figure 8 – Cash Flow Timeline of a Standard Arithmetically Decreasing Annuity

which is associated with the following terminal value:

$$(Ds)_{\overline{n}|i} = \frac{n\,(1+i)^n - s_{\overline{n}|i}}{i}. \tag{109}$$

It is clearly not financially meaningful to construct perpetuities based on decreasing annuities.

We can also directly compute the present value of an arithmetically decreasing annuity as a sum of discounted cash flows. From the cash flow timeline, it is clear that

$$(Da)_{\overline{n}|i} = \sum_{k=1}^{n}(n-k+1)\,v^k. \tag{110}$$

Eq. (110) allows us to construct a useful link between the prices of decreasing, increasing, and standard annuities. Indeed, we have [6]:

$$(Da)_{\overline{n}|i} = (n+1)\,a_{\overline{n}|i} - (Ia)_{\overline{n}|i}. \tag{112}$$

An alternative useful expression relates the prices of decreas-

---

6. Reexpress Eq. (110) as follows:

$$(Da)_{\overline{n}|i} = (n+1)\sum_{k=1}^{n} v^k - \sum_{k=1}^{n} k\,v^k. \tag{111}$$

and identify.

ing, increasing, and standard annuities. We have [7]:

$$(Da)_{\overline{n}|i} = n\, a_{\overline{n+1}|i} - v\, (Ia)_{\overline{n}|i}. \qquad (116)$$

I suggest that you check Eqs (112) and (116) using cash flow timelines.

3.2.2 Annuities and Perpetuities Due

Let us now examine the situation, illustrated in Fig. 9, where a first payment of 1 is made at time 0, payments sequentially increase by one currency unit, and a last payment of n is made at time $n-1$. This **standard arithmetically increasing annuity due** is priced as follows at time 0:

$$(I\ddot{a})_{\overline{n}|i} = \sum_{k=0}^{n-1} k\, v^k, \qquad (117)$$

so that

$$(I\ddot{a})_{\overline{n}|i} = (1+i) \cdot (Ia)_{\overline{n}|i} = \frac{\ddot{a}_{\overline{n}|i} - n\, v^n}{d}, \qquad (118)$$

using Eq. (20).

By multiplying Eq. (118) with $(1+i)^n$, we obtain the future value of this arithmetically increasing annuity due:

$$(I\ddot{s})_{\overline{n}|i} = \frac{\ddot{s}_{\overline{n}|i} - n}{d}. \qquad (119)$$

---

7. To derive this result, observe that Eq. (110) can also be written as

$$(Da)_{\overline{n}|i} = \sum_{k=1}^{n+1} (n - k + 1)\, v^k. \qquad (113)$$

by inserting a null cash flow into the sum. Then,

$$(Da)_{\overline{n}|i} = n \sum_{k=1}^{n+1} v^k - \sum_{k=1}^{n+1} (k-1)\, v^k = n \sum_{k=1}^{n+1} v^k - \sum_{k=0}^{n} k\, v^{k+1}. \qquad (114)$$

Finally,

$$(Da)_{\overline{n}|i} = n \sum_{k=1}^{n+1} v^k - v \sum_{k=1}^{n} k\, v^k, \qquad (115)$$

which is nothing but Eq. (116).

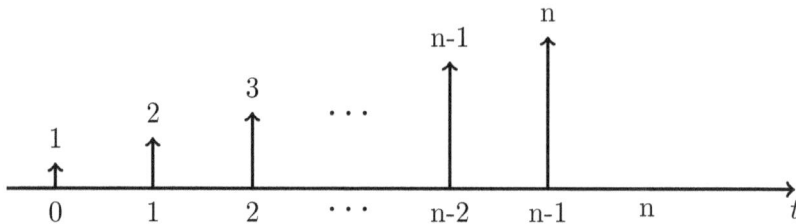

Figure 9 – Cash Flow Timeline of a Standard Arithmetically Increasing Annuity Due

Then, for a **standard arithmetically increasing perpetuity due**, we have:

$$(I\ddot{a})_{\overline{\infty}|i} = (1+i) \cdot (Ia)_{\overline{\infty}|i} = \left(\frac{1+i}{i}\right)^2 = \frac{1}{d^2}, \qquad (120)$$

where we use Eqs (107), (118), and (20) to derive this result.

A few additional important results are as follows. We first observe that, as for other similar financial products, the present value of an arithmetically increasing annuity due is equal to the present value of an otherwise identical arithmetically increasing annuity immediate multiplied by the interest factor:

$$(I\ddot{a})_{\overline{n}|i} = (1+i) \cdot (Ia)_{\overline{n}|i}, \qquad (121)$$

where this equation is analogous to Eq. (71).

Then, by analogy with Eq. (72), we write:

$$(I\ddot{a})_{\overline{n}|i} = \ddot{a}_{\overline{n}|i} + (Ia)_{\overline{n-1}|i}, \qquad (122)$$

where by separating out a sequence of unitary payments (an annuity due) from an arithmetically increasing annuity due, we recover an arithmetically increasing annuity immediate with one less cash flow. You should draw a cash flow timeline to check this fact.

By combining Eqs (121), (122), and (71), we are able to derive the following useful result:

$$(Ia)_{\overline{n}|i} = a_{\overline{n}|i} + v \, (Ia)_{\overline{n-1}|i}, \qquad (123)$$

which can be used to compute the values of arithmetically increasing annuities immediate by induction. Note that this equation can also be directly derived by drawing a cash flow timeline.

3.2.3 Several Payments per Year

Let us first consider an annuity immediate that presents mixed features: on the one hand this annuity makes m equal payments during a year ; on the other hand the payments increase year after year. The term of the annuity is n years. The m^{thly} payments are equal to $\frac{1}{m}$ in the first year, to $\frac{2}{m}$ in the second year, and so on until the last year in which they are worth $\frac{n}{m}$.

See Fig. 10 for a cash flow timeline. From the figure, we deduce that the present value of this annuity immediate, which we denote as $(Ia)_{\overline{n}|i}^{(m)}$, is equal to

$$(Ia)_{\overline{n}|i}^{(m)} = \sum_{k=1}^{n} \sum_{l=1}^{m} \frac{k}{m} v^{k-1+\frac{l}{m}}. \qquad (124)$$

We can simplify the double sum in Eq. (124) and obtain [8]:

$$(Ia)_{\overline{n}|i}^{(m)} = \frac{\ddot{a}_{\overline{n}|i} - n\, v^n}{i^{(m)}}. \qquad (128)$$

---

8. First, we write:

$$(Ia)_{\overline{n}|i}^{(m)} = \sum_{k=1}^{n} \frac{k}{m} v^{k-1} \sum_{l=1}^{m} v^{\frac{l}{m}} = \sum_{k=1}^{n} \frac{k}{m} v^{k-1} \frac{v^{\frac{1}{m}} - v^{\frac{m+1}{m}}}{1 - v^{\frac{1}{m}}}, \qquad (125)$$

where we use Eq. (3). Then, we write:

$$(Ia)_{\overline{n}|i}^{(m)} = \frac{1-v}{m\left(v^{-\frac{1}{m}} - 1\right)} \sum_{k=1}^{n} k\, v^{k-1} = \frac{1-v}{i^{(m)}} \frac{1 - (n+1)\, v^n + n\, v^{n+1}}{(1-v)^2}, \qquad (126)$$

where we recognize the expression of $i^{(m)}$ provided in Table 1 and where we use Eq. (5). Finally, we reorder terms in order to have:

$$(Ia)_{\overline{n}|i}^{(m)} = \frac{\frac{1-v^n}{1-v} - n\, v^n}{i^{(m)}} = \frac{\frac{1-v^n}{d} - n\, v^n}{i^{(m)}}, \qquad (127)$$

which is our result.

As before, by multiplying by $(1+i)^n$ the present value of this annuity immediate, we obtain its future value:

$$(Is)_{\overline{n}|i}^{(m)} = \frac{\ddot{s}_{\overline{n}|i} - n}{i^{(m)}}. \qquad (129)$$

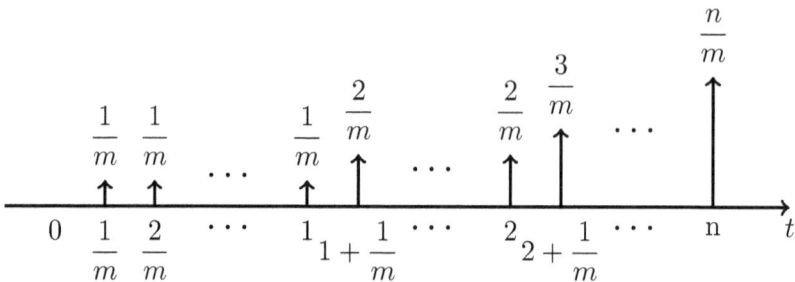

Figure 10 – Cash Flow Timeline Associated with $(Ia)_{\overline{n}|i}^{(m)}$

Taking the limit $n \to +\infty$ in Eq. (128), we can derive the present value of the corresponding perpetuity immediate:

$$(Ia)_{\overline{\infty}|i}^{(m)} = \frac{1}{i^{(m)} d}, \qquad (130)$$

where $\lim_{n \to +\infty} \ddot{a}_{\overline{n}|i} = \frac{1}{d}$.

Next, we consider a financial product that has the same characteristics as above, except that it is an annuity due. Thus, this financial product makes m payments of $\frac{1}{m}$ at times $0, \frac{1}{m}, \frac{2}{m}, \cdots$, m payments of $\frac{2}{m}$ at times $1, 1+\frac{1}{m}, 1+\frac{2}{m}, \cdots$, and so on until the last series of m payments of $\frac{n}{m}$ that are made at times $n-1, n-1+\frac{1}{m}, \cdots, n-1+\frac{m-1}{m} = n - \frac{1}{m}$. No payment is made at time n. See Fig. 11 for a cash flow timeline.

The present value of this annuity due is computed by multiplying by $\left(1 + \frac{i^{(m)}}{m}\right)$ the present value of the otherwise equivalent annuity immediate that is provided in Eq. (128). Indeed, in the spirit of Eq. (71), the present value of an annuity due is equal to the present value of an otherwise equivalent annuity immediate multiplied by an interest factor that is computed based on the time lag between payments.

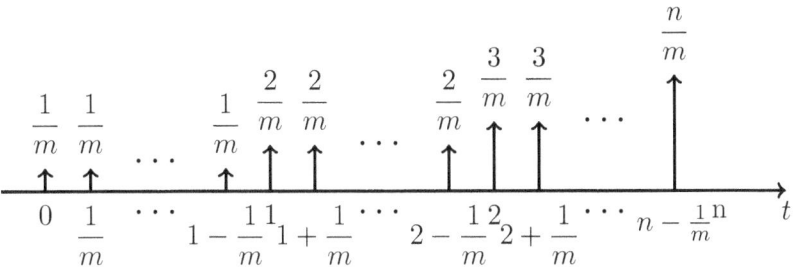

Figure 11 – Cash Flow Timeline Associated with $(I\ddot{a})^{(m)}_{\overline{n}|i}$

Therefore, the present value $(I\ddot{a})^{(m)}_{\overline{n}|i}$ of this annuity due is equal to

$$(I\ddot{a})^{(m)}_{\overline{n}|i} = \frac{\ddot{a}_{\overline{n}|i} - n\, v^n}{d^{(m)}}, \tag{131}$$

where we use the fact that $d^{(m)} = \frac{i^{(m)}}{\left(1+\frac{i^{(m)}}{m}\right)}$. Then, by multiplying Eq. (131) by $(1+i)^n$, we deduce the following future value:

$$(I\ddot{s})^{(m)}_{\overline{n}|i} = \frac{\ddot{s}_{\overline{n}|i} - n}{d^{(m)}}. \tag{132}$$

We observe again that $\lim_{n \to +\infty} \ddot{a}_{\overline{n}|i} = \frac{1}{d}$. The present value of the corresponding perpetuity due follows:

$$(I\ddot{a})^{(m)}_{\overline{\infty}|i} = \frac{1}{d^{(m)}\, d}. \tag{133}$$

Let us now study a different class of increasing annuities making m payments per year. For this new class of financial products, payments do not increase every year. Instead, they increase one after the other. Let the payment times be $\frac{1}{m}, \frac{2}{m}, \cdots, n - \frac{1}{m}, n$, where we deal with an annuity immediate. Then, the payment amounts are assumed to be equal to the payment times divided by m. Specifically, the payment amounts are equal to $\frac{1}{m^2}, \frac{2}{m^2}, \cdots, \frac{n}{m} - \frac{1}{m^2}, \frac{n}{m}$. See Fig. 12 for a cash flow timeline.

We first derive a series expression, denoted by $(I^{(m)}s)^{(m)}_{\overline{n}|i}$, for

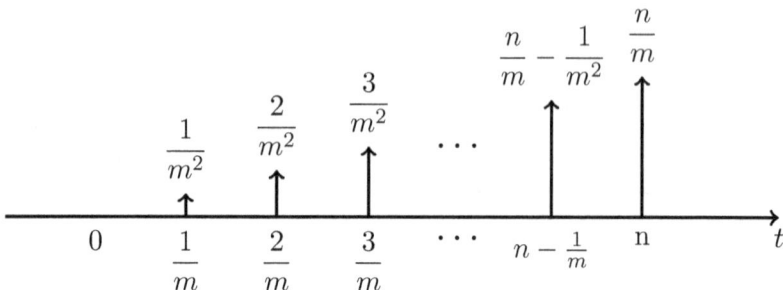

Figure 12 – Cash Flow Timeline Associated with $(I^{(m)}a)^{(m)}_{\overline{n}|i}$

the present value of this financial product. We write:

$$(I^{(m)}s)^{(m)}_{\overline{n}|i} = \sum_{k=1}^{nm} \frac{k}{m^2} v^{\frac{k}{m}}, \qquad (134)$$

where the payment times are equal to $\frac{k}{m}$, the payment amounts to $\frac{k}{m^2}$, and nm payments are made in total.

The series expression in Eq. (134) can be simplified a lot [9]. We obtain:

$$(I^{(m)}a)^{(m)}_{\overline{n}|i} = \frac{\ddot{a}^{(m)}_{\overline{n}|i} - n\, v^n}{i^{(m)}}, \qquad (137)$$

which is associated with the following future value:

$$(I^{(m)}s)^{(m)}_{\overline{n}|i} = \frac{\ddot{s}^{(m)}_{\overline{n}|i} - n}{i^{(m)}}. \qquad (138)$$

---

9. Hint:

$$(I^{(m)}s)^{(m)}_{\overline{n}|i} = \frac{v^{\frac{1}{m}}}{m^2} \sum_{k=1}^{nm} k \left(v^{\frac{1}{m}}\right)^{k-1} \qquad (135)$$

yields

$$(I^{(m)}s)^{(m)}_{\overline{n}|i} = \frac{v^{\frac{1}{m}}}{m^2} \frac{1 - (nm+1)\left(v^{\frac{1}{m}}\right)^{nm} + nm\left(v^{\frac{1}{m}}\right)^{nm+1}}{\left(1 - v^{\frac{1}{m}}\right)^2}, \qquad (136)$$

using Eq. (5). Then, reorder terms and identify $i^{(m)} = m\, v^{-\frac{1}{m}}\left(1 - v^{\frac{1}{m}}\right)$, $d^{(m)} = m\left(1 - v^{\frac{1}{m}}\right)$, and $\ddot{a}^{(m)}_{\overline{n}|i} = \frac{1 - v^n}{d^{(m)}}$ to conclude.

The present value of the corresponding perpetuity is equal to

$$(I^{(m)}a)^{(m)}_{\overline{\infty}|i} = \frac{1}{i^{(m)}\, d^{(m)}}, \qquad (139)$$

where we use the fact that $\lim_{n\to+\infty} \ddot{a}^{(m)}_{\overline{n}|i} = \frac{1}{d^{(m)}}$.

Let us now consider an annuity due whose payment times are equal to $0, \frac{1}{m}, \frac{2}{m}, \cdots, n-\frac{1}{m}$ and whose payment amounts are equal to $\frac{1}{m^2}, \frac{2}{m^2}, \cdots, \frac{n}{m} - \frac{1}{m^2}, \frac{n}{m}$. See Fig. 13 for a cash flow timeline.

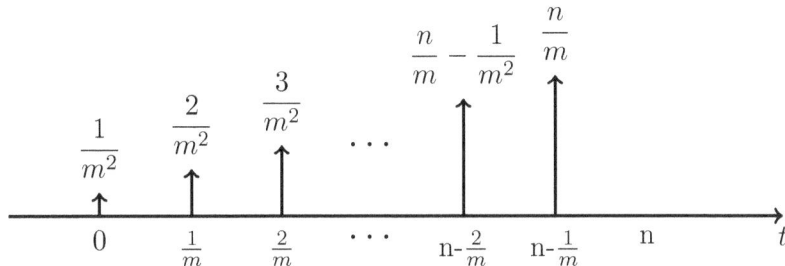

Figure 13 – Cash Flow Timeline Associated with $(I^{(m)}\ddot{a})^{(m)}_{\overline{n}|i}$

Reasoning as before, we obtain the present value $(I^{(m)}\ddot{a})^{(m)}_{\overline{n}|i}$ of this annuity due by multiplying by $\left(1+\frac{i^{(m)}}{m}\right)$ the present value $(I^{(m)}a)^{(m)}_{\overline{n}|i}$ of the otherwise equivalent annuity immediate. Therefore,

$$(I^{(m)}\ddot{a})^{(m)}_{\overline{n}|i} = \frac{\ddot{a}^{(m)}_{\overline{n}|i} - n\, v^n}{d^{(m)}}, \qquad (140)$$

where again we use the fact that $d^{(m)} = \frac{i^{(m)}}{\left(1+\frac{i^{(m)}}{m}\right)}$. Then, by multiplying Eq. (140) by $(1+i)^n$, we deduce the following future value:

$$(I^{(m)}\ddot{s})^{(m)}_{\overline{n}|i} = \frac{\ddot{s}^{(m)}_{\overline{n}|i} - n}{d^{(m)}}. \qquad (141)$$

Taking the limit $n \to +\infty$ in Eq. (140), we can derive the present value of the corresponding perpetuity immediate:

$$(I^{(m)}\ddot{a})^{(m)}_{\overline{\infty}|i} = \frac{1}{\left(d^{(m)}\right)^2}, \qquad (142)$$

where $\lim_{n \to +\infty} \ddot{a}^{(m)}_{\overline{n}|i} = \frac{1}{d^{(m)}}$.

3.2.4 Continuous Annuities and Perpetuities

Continuous arithmetically increasing annuities and perpetuities do not make discrete payments. Instead, they provide a continuous rate of payment that increases every year.

First, we consider the case of an increasing annuity that offers a rate of payment of 1 during period $(0,1)$, of 2 during period $(1,2)$, of 3 during period $(2,3)$, and so on. Therefore, this annuity provides a rate of payment of k during each period $(k-1, k)$. However, the discount factor that should be used at each point of time t of period $(k-1, k)$ is distinct and equal to v^t. Therefore, the present value of receiving a rate of payment of k during period $(k-1, k)$ is worth $k \int_{k-1}^{k} v^t \, dt$.

Summing up the present values of the contributions made over several periods, we obtain the present value of an increasing annuity that makes continuous payments [10]:

$$(I\bar{a})_{\overline{n}|i} = \sum_{k=1}^{n} k \int_{k-1}^{k} v^t \, dt = \frac{\ddot{a}_{\overline{n}|i} - n\, v^n}{\delta}. \qquad (145)$$

---

10. Hint: compute the integrals and show that

$$\sum_{k=1}^{n} k \int_{k-1}^{k} v^t \, dt = \frac{1-v}{\delta} \sum_{k=1}^{n} k\, v^{k-1}, \qquad (143)$$

where you should check that the primitive function of v^t is $-\frac{v^t}{\delta}$. Then, using Eq. (5) allows you to write:

$$\sum_{k=1}^{n} k \int_{k-1}^{k} v^t \, dt = \frac{1-v^n}{d\,\delta} - \frac{n\,v^n}{\delta}. \qquad (144)$$

Recognizing $\ddot{a}_{\overline{n}|i}$ allows you to conclude. Again, I advise you to check all of these steps on a separate sheet of paper: it will strengthen your mastery of the relations between v, d and δ.

The future value of this continuous annuity readily follows by compounding the previous result with $(1+i)^n$:

$$(I\bar{s})_{\overline{n}|i} = \frac{\ddot{s}_{\overline{n}|i} - n}{\delta}. \tag{146}$$

Next, the present value of an increasing perpetuity that makes continuous payments can be derived by taking the limit $n \to +\infty$ in Eq. (145). We obtain:

$$(I\bar{a})_{\overline{\infty}|i} = \frac{\ddot{a}_{\overline{\infty}|i}}{\delta} = \frac{1}{\delta\, d}, \tag{147}$$

where we use the fact that $\lim_{n \to +\infty} n\, v^n = 0$ and we recognize $\ddot{a}_{\overline{\infty}|i} = \frac{1}{d}$.

Now, not only do we assume that the annuity provides a continuous rate of payment, but we also postulate that this rate of payment increases in a continuous way. Therefore, we assume that a rate of payment of t is made at any point of time t until time T. This continuous rate of payment is associated with a present value of $t\, v^t$. By adapting Eq. (145), we are able to compute the present value $(\bar{I}\bar{a})_{\overline{T}|i}$ of an annuity that makes continuous payments and whose continuous payments continuously increase [11]:

$$(\bar{I}\bar{a})_{\overline{T}|i} = \int_0^T t\, v^t\, dt = \frac{\bar{a}_{\overline{T}|i} - T\, v^T}{\delta}. \tag{148}$$

The future value at time T of this annuity is computed by multiplying $(\bar{I}\bar{a})_{\overline{T}|i}$ by $(1+i)^T$. We obtain:

$$(\bar{I}\bar{s})_{\overline{T}|i} = \frac{\bar{s}_{\overline{T}|i} - T}{\delta}. \tag{149}$$

Next, the present value of a perpetuity that makes continuous payments that continuously increase can be derived by taking the limit $T \to +\infty$ in Eq. (148). We obtain:

$$(\bar{I}\bar{a})_{\overline{\infty}|i} = \frac{\bar{a}_{\overline{\infty}|i}}{\delta} = \frac{1}{\delta^2}, \tag{150}$$

---

11. Hint: perfom an integration by parts and use the fact that the primitive function of v^t is $-\frac{v^t}{\delta}$. Use Eq. (91) to conclude.

where we use the fact that $\lim_{T\to+\infty} T\, v^T = 0$ and we recognize $\bar{a}_{\overline{\infty}|i} = \frac{1}{\delta}$.

3.2.5 Conclusion

We show in Table 3 a summary of the present and future values of arithmetically increasing annuities and of the present values of arithmetically increasing perpetuities.

Unfortunately, it is not possible to rank the present and future values of arithmetically increasing annuities in a systematic and consistent way.

However, it is possible to rank the present values of arithmetically increasing perpetuities. For this purpose, we recall that

$$d < d^{(m)} < \delta < i^{(m)} < i, \qquad (151)$$

from which you can deduce inequalities such as $d^{(m)}d > d^2$.

You can also check in a spreadsheet [12] that the following more sophisticated inequalities hold for realistic interest rate values:

$$id > i^{(m)}\, d^{(m)} > \delta^2 \qquad (152)$$

and

$$\left(d^{(m)}\right)^2 > i^{(m)}\, d. \qquad (153)$$

These inequalities allow us to rank the present values of perpetuities as follows:

$$(Ia)_{\overline{\infty}|i} < (I^{(m)}a)^{(m)}_{\overline{\infty}|i} < (I\bar{a})_{\overline{\infty}|i} < (I^{(m)}\ddot{a})^{(m)}_{\overline{\infty}|i} < \cdots \qquad (154)$$

and

$$\cdots < (Ia)^{(m)}_{\overline{\infty}|i} < (I\bar{a})_{\overline{\infty}|i} < (I\ddot{a})^{(m)}_{\overline{\infty}|i} < (I\ddot{a})_{\overline{\infty}|i}. \qquad (155)$$

3.3 Geometric Growth

Geometric growth differentiates itself from arithmetic growth by payments that increase by a certain rate rather than by a certain amount.

---

12. An analytical proof seems to be a challenge.

$(Ia)_{\overline{n\vert}i}$	$(I^{(m)}a)^{(m)}_{\overline{n\vert}i}$	$(\bar{I}\bar{a})_{\overline{T\vert}i}$	$(I^{(m)}\ddot{a})^{(m)}_{\overline{n\vert}i}$	$(Ia)^{(m)}_{\overline{n\vert}i}$	$(I\bar{a})_{\overline{n\vert}i}$	$(I\ddot{a})^{(m)}_{\overline{n\vert}i}$	$(I\ddot{a})_{\overline{n\vert}i}$
$\frac{\ddot{a}_{\overline{n\vert}i}-n\ v^n}{i}$	$\frac{\ddot{a}^{(m)}_{\overline{n\vert}i}-n\ v^n}{i^{(m)}}$	$\frac{\bar{a}_{\overline{T\vert}i}-T\ v^T}{\delta}$	$\frac{\ddot{a}^{(m)}_{\overline{n\vert}i}-n\ v^n}{d^{(m)}}$	$\frac{\ddot{a}_{\overline{n\vert}i}-n\ v^n}{i^{(m)}}$	$\frac{\ddot{a}_{\overline{n\vert}i}-n\ v^n}{\delta}$	$\frac{\ddot{a}_{\overline{n\vert}i}-n\ v^n}{d^{(m)}}$	$\frac{\ddot{a}_{\overline{n\vert}i}-n\ v^n}{d}$
109.292	106.002	105.696	106.445	111.837	112.070	112.304	114.895
$(Is)_{\overline{n\vert}i}$	$(I^{(m)}s)^{(m)}_{\overline{n\vert}i}$	$(\bar{I}\bar{s})_{\overline{T\vert}i}$	$(I^{(m)}\ddot{s})^{(m)}_{\overline{n\vert}i}$	$(Is)^{(m)}_{\overline{n\vert}i}$	$(I\bar{s})_{\overline{n\vert}i}$	$(I\ddot{s})^{(m)}_{\overline{n\vert}i}$	$(I\ddot{s})_{\overline{n\vert}i}$
$\frac{\ddot{s}_{\overline{n\vert}i}-n}{i}$	$\frac{\ddot{s}^{(m)}_{\overline{n\vert}i}-n}{i^{(m)}}$	$\frac{\bar{s}_{\overline{T\vert}i}-T}{\delta}$	$\frac{\ddot{s}^{(m)}_{\overline{n\vert}i}-n}{d^{(m)}}$	$\frac{\ddot{s}_{\overline{n\vert}i}-n}{i^{(m)}}$	$\frac{\ddot{s}_{\overline{n\vert}i}-n}{\delta}$	$\frac{\ddot{s}_{\overline{n\vert}i}-n}{d^{(m)}}$	$\frac{\ddot{s}_{\overline{n\vert}i}-n}{d}$
297.086	288.144	287.313	289.348	304.005	304.639	305.274	312.318
$(Ia)_{\overline{\infty\vert}i}$	$(I^{(m)}a)^{(m)}_{\overline{\infty\vert}i}$	$(\bar{I}\bar{a})_{\overline{\infty\vert}i}$	$(I^{(m)}\ddot{a})^{(m)}_{\overline{\infty\vert}i}$	$(Ia)^{(m)}_{\overline{\infty\vert}i}$	$(I\bar{a})_{\overline{\infty\vert}i}$	$(I\ddot{a})^{(m)}_{\overline{\infty\vert}i}$	$(I\ddot{a})_{\overline{\infty\vert}i}$
$\frac{1}{i\ d}$	$\frac{1}{i^{(m)}\ d^{(m)}}$	$\frac{1}{\delta^2}$	$\frac{1}{\left(d^{(m)}\right)^2}$	$\frac{1}{i^{(m)}\ d}$	$\frac{1}{\delta\ d}$	$\frac{1}{d^{(m)}\ d}$	$\frac{1}{d^2}$
399.917	399.999	400.000	401.670	409.230	410.083	410.938	420.421

Table 3 – Valuation of Arithmetically Increasing Annuities and Perpetuities. The Illustrations are Computed Using $T = n = 20$, $m = 12$, and $\delta = 5\%$.

3.3.1 Annuities and Perpetuities Immediate

We start by computing the **present value of a geometrically increasing annuity** that makes n payments growing at a rate of g. We assume that the first payment of this annuity is equal to 1 and is made at time 1. Therefore, its second payment is equal to $1+g$ and is made at time 2, its third payment is equal to $(1+g)^2$ and is made at time 3, and so on until its last payment of $(1+g)^{n-1}$ that is made at time n [13]. See Fig. 14 for a cash flow timeline. As before, the effective interest rate is i. The price of this annuity is [14]

$$(Ga)_{\overline{n}|i,g} = \frac{v - (1+g)^n v^{n+1}}{1 - (1+g)v} = \frac{1 - \left(\frac{1+g}{1+i}\right)^n}{i - g}, \quad (156)$$

when $g < i$ or $g > i$. This formula also holds for a geometrically decreasing annuity, so when $g < 0$.

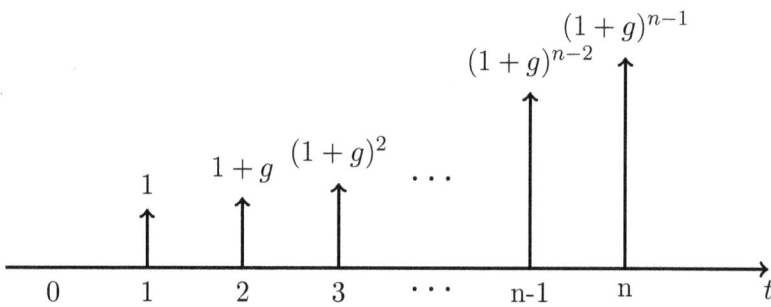

Figure 14 – Cash Flow Timeline of a Geometrically Increasing Annuity Immediate

---

13. Pay a particular attention to the value of the first payment and to its time. Modifications to these initial conditions will be made in the exercises and will radically change the price of the product.

14. To show the first equality, compute $(Ga)_{\overline{n}|i,g} = \sum_{k=1}^{n}(1+g)^{k-1}v^k = v\sum_{k=0}^{n-1}(1+g)^k v^k$ using the geometric series result in Eq. (1). To show the second equality, factor out v.

An excellent exercise consists in checking that

$$(Ga)_{\overline{n}|i,g} = \frac{1}{1+g} a_{\overline{n}|j} \qquad (157)$$

and

$$(Ga)_{\overline{n}|i,g} = \frac{1}{1+i} \ddot{a}_{\overline{n}|j}, \qquad (158)$$

where we define the new rate [15]:

$$j = \frac{i-g}{1+g}. \qquad (159)$$

Note that j allows us to construct an interest factor:

$$1+j = \frac{1+i}{1+g} \qquad (160)$$

and a discount factor

$$v_j = \frac{1}{1+j} = \frac{1+g}{1+i} \qquad (161)$$

that are often useful when we price geometric annuities of different kinds.

When payments increase at a rate equal to the effective interest rate, so when $g = i$, the present value of the geometrically increasing annuity is simplified a lot and becomes:

$$(Ga)_{\overline{n}|i,i} = \frac{n}{1+i}. \qquad (162)$$

The **future value of a geometrically increasing annuity** is worth

$$(Gs)_{\overline{n}|i,g} = (1+i)^n \frac{1 - \left(\frac{1+g}{1+i}\right)^n}{i-g} = \frac{(1+i)^n - (1+g)^n}{i-g}, \qquad (163)$$

when $g < i$ or $g > i$, while we have:

$$(Gs)_{\overline{n}|i,i} = n \cdot (1+i)^{n-1}, \qquad (164)$$

---

[15]. Be careful that the rate j defined here is not a semiannual effective interest rate. I kept this notation because you will encounter it a lot elsewhere.

when $g = i$.

Let us now examine the case of an infinite horizon. The **present value of a geometrically increasing perpetuity immediate**, whose payments grow at a rate of $g < i$, is

$$(Ga)_{\overline{\infty}|i,g} = \frac{1}{i-g} = \frac{v}{1 - v(1+g)}, \qquad (165)$$

where the first payment is 1 at time 1, the second payment is $1 + g$ at time 2, the third payment is $(1+g)^2$ at time 3, and so on. This formula is derived by letting $n \to +\infty$ in Eq. (156) and it only holds when $g < i$ because this is the only situation where $\left(\frac{1+g}{1+i}\right)^n$ converges to 0.

3.3.2 Annuities and Perpetuities Due

We now consider the case of a geometrically increasing annuity due, whose first payment of 1 occurs at time 0. Again, we assume that n payments are made and that these payments grow at a rate of g. See Fig. 15 for a cash flow timeline.

We can readily compute the present value of a geometrically increasing annuity due as follows:

$$(G\ddot{a})_{\overline{n}|i,g} = (1+i) \cdot (Ga)_{\overline{n}|i,g} = \frac{1 - (1+g)^n v^n}{1 - (1+g) v} \qquad (166)$$

when $g < i$ or $g > i$, and

$$(G\ddot{a})_{\overline{n}|i,i} = (1+i) \cdot (Ga)_{\overline{n}|i,i} = n \qquad (167)$$

when $g = i$.

You can also check that

$$(G\ddot{a})_{\overline{n}|i,g} = (1+j) \, a_{\overline{n}|j} \qquad (168)$$

and

$$(G\ddot{a})_{\overline{n}|i,g} = \ddot{a}_{\overline{n}|j}, \qquad (169)$$

when we define again $j = \frac{i-g}{1+g}$.

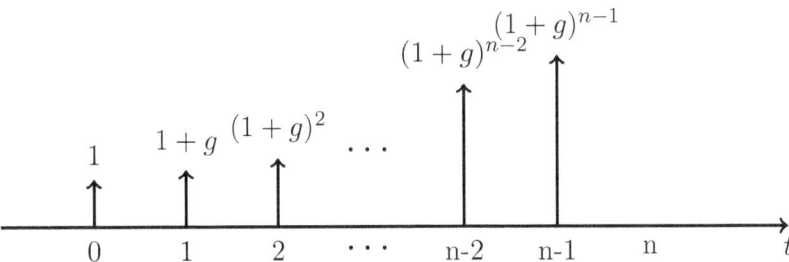

Figure 15 – Cash Flow Timeline of a Geometrically Increasing Annuity Due

The **present value of a geometrically increasing perpetuity due**, whose payments grow at a rate of $g < i$, is

$$(G\ddot{a})_{\overline{\infty}|i,g} = (1+i) \cdot (Ga)_{\overline{\infty}|i,g} = \frac{1+i}{i-g} = \frac{1}{1-v(1+g)}, \quad (170)$$

where the first payment is 1 at time 0, the second payment is $1+g$ at time 1, the third payment is $(1+g)^2$ at time 2, and so on.

3.3.3 Special Cases

In some exercises, a reference amount of 1 is given at time 0, but the first payment is made at time 1. This is equivalent to valuing a growing annuity with a first payment of $1+g$ at time 1, n payments, an effective interest rate of i, and a growth rate of g. For such a financial product, we compute the following present value:

$$(1+g) \cdot (Ga)_{\overline{n}|i,g} = (1+g) \frac{1 - \left(\frac{1+g}{1+i}\right)^n}{i-g} \quad (171)$$

when $g < i$ or $g > i$.

You can derive Eq. (171) by computing $\sum_{k=1}^{n} [(1+g) \, v]^k$. Clearly, Eq. (156) can be recovered by dividing Eq. (171) by $1+g$.

For a geometrically increasing perpetuity that makes a first payment of $1+g$ at time 1, a second payment of $(1+g)^2$ at time 2, a third payment of $(1+g)^3$ at time 3, and so on, the pricing formula becomes:

$$(1+g) \cdot (Ga)_{\overline{\infty}|i,g} = \frac{1+g}{i-g}, \qquad (172)$$

where we take the limit $n \to +\infty$ in Eq. (171).

If now we consider a perpetuity that makes a first payment of $1+g$ at time 0, a second payment of $(1+g)^2$ at time 1, a third payment of $(1+g)^3$ at time 2, and so on, we compute the following present value:

$$(1+g) \cdot (G\ddot{a})_{\overline{\infty}|i,g} = \frac{1+g}{1-v(1+g)}. \qquad (173)$$

The pricing of analogous geometrically increasing annuities and perpetuities follows in a similar way.

3.3.4 Continuous Annuities and Perpetuities

To conclude this section, we note that the **present value of a geometrically increasing continuous annuity** is equal to [16]

$$(G\overline{a})_{\overline{t}|i} = \int_0^t (1+g)^s v^s ds = \frac{1-(1+g)^t v^t}{\ln(1+i) - \ln(1+g)}, \qquad (174)$$

where we recover the present value of a standard non-increasing continuous annuity (shown in Eq. (91)) when $g = 0$.

We can also compute the **present value of a geometrically increasing continuous perpetuity**. We obtain:

$$(G\overline{a})_{\overline{\infty}|i,g} = \int_0^{+\infty} (1+g)^s v^s ds = \frac{1}{\ln(1+i) - \ln(1+g)}, \qquad (175)$$

where we recover the present value of a standard non-increasing continuous perpetuity (shown in Eq. (93)) when $g = 0$.

---

[16]. Hint: check that the primitive function of $(1+g)^s v^s$ is $\frac{(1+g)^s v^s}{\ln(1+g) - \ln(1+i)}$ and use this property.

3.4 A Few Bonus Results

Let us conclude this chapter with some useful complementary results on annuities and perpetuities.

3.4.1 Inverse Values

It is possible to provide a simple link between $s_{\overline{n}|i}$ and $a_{\overline{n}|i}$, which, contrary to Eq. (67), only depends on i and does not require computing a power of the discount factor. Indeed, we have [17]:

$$\frac{1}{a_{\overline{n}|i}} = \frac{1}{s_{\overline{n}|i}} + i. \qquad (177)$$

Similarly, $\ddot{s}_{\overline{n}|i}$ and $\ddot{a}_{\overline{n}|i}$ are related in a very simple way through their reference rate (d) as follows:

$$\frac{1}{\ddot{a}_{\overline{n}|i}} = \frac{1}{\ddot{s}_{\overline{n}|i}} + d. \qquad (178)$$

We also have:

$$\frac{1}{\ddot{a}_{\overline{n}|i}^{(m)}} = \frac{1}{\ddot{s}_{\overline{n}|i}^{(m)}} + d^{(m)} \qquad (179)$$

and

$$\frac{1}{a_{\overline{n}|i}^{(m)}} = \frac{1}{s_{\overline{n}|i}^{(m)}} + i^{(m)} \qquad (180)$$

and also

$$\frac{1}{\bar{a}_{\overline{t}|i}} = \frac{1}{\bar{s}_{\overline{t}|i}} + \delta. \qquad (181)$$

---

[17]. This result is a consequence of:

$$1 = v^n + (1 - v^n) = \frac{a_{\overline{n}|i}}{s_{\overline{n}|i}} + i\frac{1 - v^n}{i} = a_{\overline{n}|i}\left[\frac{1}{s_{\overline{n}|i}} + i\right], \qquad (176)$$

where we use Eqs (64) and (67). The proofs of the next results follow by replacing i with d, $d^{(m)}$, $i^{(m)}$, and δ in the above series of equalities.

3.4.2 Less Than a Payment per Year

Let us now examine the valuation of an annuity *immediate* whose payments are made every m years and that makes h payments in total. The term of this annuity is equal to $n = mh$. The present value of this annuity immediate can be computed as follows:

$$\sum_{k=1}^{h}(v^m)^k = \frac{v^m - v^{m(h+1)}}{1 - v^m}, \qquad (182)$$

using Eq. (3). The future value at time n of this annuity immediate is equal to

$$\sum_{k=0}^{h-1}((1+i)^m)^k = \frac{1 - (1+i)^{mh}}{1 - (1+i)^m}, \qquad (183)$$

where we use Eq. (1). Alternatively, we can factor out v^m in both the numerator and the denominator of Eq. (182) and compound this equation by $(1+i)^{mh}$ to obtain Eq. (183).

In Eq. (183), $k = 0$ corresponds to the last payment that should not be compounded because it already occurs at time $n = mh$, $k = 1$ corresponds to the penultimate payment that should be compounded over m years, $k = 2$ corresponds to the antepenultimate payment that should be compounded over $2m$ years, and so on for higher values of k.

By extension, we compute the present value of a perpetuity *immediate* that pays one currency unit every m years as follows:

$$\sum_{k=1}^{\infty}(v^m)^k = \frac{v^m}{1 - v^m} = \frac{1}{(1+i)^m - 1}, \qquad (184)$$

where we use Eq. (4). This result can also be derived by taking the limit $h \to +\infty$ in Eq. (182).

We now examine the valuation of an annuity *due* whose payments are made every m years and that makes h payments in total. The term of this annuity is again equal to $n = h \cdot m$. The present value of this annuity due can be computed as follows:

$$\sum_{k=0}^{h-1}(v^m)^k = \frac{1 - v^{mh}}{1 - v^m}, \qquad (185)$$

using Eq. (1). The future value at time n of this annuity due is equal to

$$\sum_{k=1}^{h} ((1+i)^m)^k = \frac{(1+i)^m - (1+i)^{m(h+1)}}{1 - (1+i)^m}, \quad (186)$$

where we use Eq. (3). Alternatively, we can factor out v^m in both the numerator and the denominator of Eq. (185) and multiply this equation by $(1+i)^{mh}$ to obtain Eq. (186).

In Eq. (186), $k = 1$ corresponds to the last payment that should be compounded from its occurrence time $m(h-1)$ to the term $n = mh$ using the compounding factor $(1+i)^m$, $k = 2$ corresponds to the penultimate payment that should be compounded from its occurrence time $m(h-2)$ to the term $n = mh$ using the compounding factor $(1+i)^{2m}$, and so on for higher values of k.

The present value of a perpetuity *due* that pays one currency unit every m years is equal to

$$\sum_{k=0}^{\infty} (v^m)^k = \frac{1}{1 - v^m} = \frac{(1+i)^m}{(1+i)^m - 1}, \quad (187)$$

where we use Eq. (2). This result can also be derived by taking the limit $h \to +\infty$ in Eq. (185).

3.4.3 Asymptotic Results

Hopefully, the results that we derived in the presence of multiple payments per year converge to the results that we derived in the presence of continuous payments when the frequency of payments becomes infinite. Indeed, using Eqs (82) and (87) together with $\lim_{m \to +\infty} i^{(m)} = \delta$ and $\lim_{m \to +\infty} d^{(m)} = \delta$, we can derive the following asymptotic result:

$$\lim_{m \to +\infty} a_{\overline{n}|i}^{(m)} = \lim_{m \to +\infty} \ddot{a}_{\overline{n}|i}^{(m)} = \bar{a}_{\overline{n}|i}. \quad (188)$$

Similarly, we have:

$$\lim_{m \to +\infty} s_{\overline{n}|i}^{(m)} = \lim_{m \to +\infty} \ddot{s}_{\overline{n}|i}^{(m)} = \bar{s}_{\overline{n}|i} \quad (189)$$

and
$$\lim_{m\to+\infty} a^{(m)}_{\overline{\infty}|i} = \lim_{m\to+\infty} \ddot{a}^{(m)}_{\overline{\infty}|i} = \bar{a}_{\overline{\infty}|i}. \qquad (190)$$

We can also use the expressions provided in Table 3 together with $\lim_{m\to+\infty} i^{(m)} = \delta$ and $\lim_{m\to+\infty} d^{(m)} = \delta$ to derive the following asymptotic results for the present values of arithmetically increasing annuities:

$$\lim_{m\to+\infty} (I^{(m)}a)^{(m)}_{\overline{n}|i} = \lim_{m\to+\infty} (I^{(m)}\ddot{a})^{(m)}_{\overline{n}|i} = (\bar{I}\bar{a})_{\overline{n}|i} \qquad (191)$$

and

$$\lim_{m\to+\infty} (Ia)^{(m)}_{\overline{n}|i} = \lim_{m\to+\infty} (I\ddot{a})^{(m)}_{\overline{n}|i} = (I\bar{a})_{\overline{n}|i}. \qquad (192)$$

Similar asymptotic results hold for the future values of arithmetically increasing annuities and for the present values of arithmetically increasing perpetuities.

3.4.4 Deferred Annuities

By definition, an annuity that is deferred m years starts making payments m years later than it would otherwise do. For example, an arithmetically increasing annuity immediate deferred m years, denoted by $_{m|}(Ia)_{\overline{n}|i}$, makes its first payment of 1 at time $m+1$ instead of at time 1. This annuity makes its last payment of n at time $n+m$ instead of at time n. See Fig. 16 for a cash flow timeline.

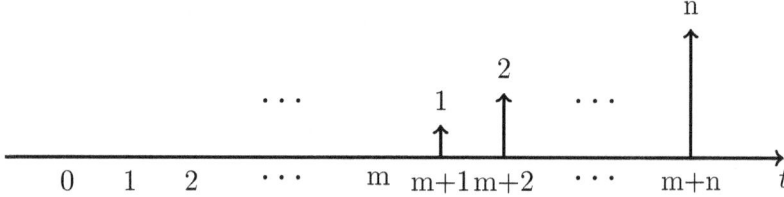

Figure 16 – Cash Flow Timeline of a Deferred Annuity Immediate

Consider for instance a standard (non-increasing) deferred annuity immediate. This annuity can be valued as follows:

$$_{m|}a_{\overline{n}|i} = v^m\, a_{\overline{n}|i}, \qquad (193)$$

where we first compute $a_{\overline{n}|i}$, which represents the present value at time m of the n deferred cash flows. Then, we discount this value from time m to time 0 with v^m to obtain the present value at time 0 of the n deferred cash flows.

Other deferred annuities, such as deferred annuities due, deferred arithmetically increasing or decreasing annuities, ..., are valued using a formula similar to (193).

Consider now a standard annuity immediate that makes $m+n$ payments in total. This annuity can be decomposed into a standard annuity immediate that makes m payments and an annuity deferred m years that makes n payments. Based on this observation, can we derive the following pricing relation:

$$a_{\overline{m+n}|i} = a_{\overline{m}|i} +_{m|} a_{\overline{n}|i}, \tag{194}$$

or, using Eq. (193),

$$a_{\overline{m+n}|i} = a_{\overline{m}|i} + v^m\, a_{\overline{n}|i}. \tag{195}$$

By multiplying Eq. (195) by $(1+i)^{m+n}$, we also obtain:

$$s_{\overline{m+n}|i} = (1+i)^n\, s_{\overline{m}|i} + s_{\overline{n}|i}, \tag{196}$$

where we use the fact that $s_{\overline{n}|i} = a_{\overline{n}|i}(1+i)^n$ for any value of n.

Eq. (196) can be interpreted as follows. First, the left-hand-side of the equation is the future value of an annuity immediate that makes $m+n$ payments. Next, the right-hand-side of the equation can be decomposed into two components. The first of these components corresponds to a future value computed at time m (of an annuity immediate that makes m payments of 1 between times 1 and m) that is compounded from time m to time $m+n$ thanks to $(1+i)^n$. The second of these components corresponds to the future value computed at time $m+n$ of n payments of 1 made between times $m+1$ and $m+n$.

Beware of this last interpretation. We have written that $s_{\overline{n}|i}$, so the future value computed at time n of n payments of 1 made between times 1 and n, is identical to the future value computed at time $m+n$ of n payments of 1 made between times $m+1$ and $m+n$. We could write that because we made a very strong and hidden assumption. Indeed, we assumed that pricing is **stationary**. According to this assumption, you can price financial products using the same recipes independently of the valuation time. This assumption will blow up when we consider life annuities, for which pricing depends on the age of the policyholder. You will need to meditate a lot on this important observation when you prepare for exam LTAM.

3.4.5 Rainbow Annuities

We now come to the valuation of a **rainbow annuity immediate** that makes $2n-1$ payments. Its first n payments increase from the value 1 at time 1 to the value n at time n. Its last $n-1$ payments decrease from the value $n-1$ at time $n+1$ to the value 1 at time $2n-1$. See Fig. 17 for a cash flow timeline.

A rainbow annuity combines an increasing annuity immediate that makes n payments and a deferred decreasing annuity immediate that makes $n-1$ payments. The increasing annuity component is worth $(Ia)_{\overline{n}|i}$ at time 0, while the decreasing annuity component is worth $(Da)_{\overline{n-1}|i}$ at time n. The latter value should be discounted to time 0 using the discount factor v^n. Thus, the price P^{RAI} of a rainbow annuity immediate is

$$P^{\text{RAI}} = (Ia)_{\overline{n}|i} + v^n (Da)_{\overline{n-1}|i}. \tag{197}$$

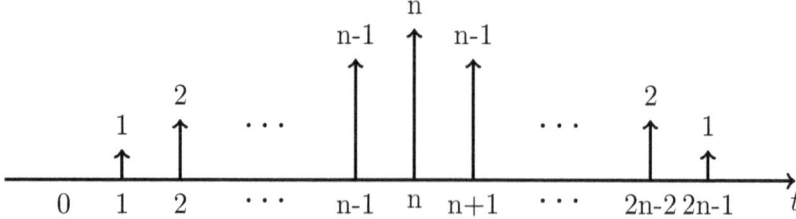

Figure 17 – Cash Flow Timeline of a Standard Rainbow Annuity Immediate

Alternatively, we can show that [18]

$$P^{\text{RAI}} = \ddot{a}_{\overline{n}|i}\, a_{\overline{n}|i}. \qquad (200)$$

A related financial product is the **paused rainbow annuity immediate**, which makes $2n$ payments in total. Its first n payments increase from the value 1 at time 1 to the value n at time n. Its last n payments decrease from the value n at time $n+1$ to the value 1 at time $2n$. Thus, the amount n is paid twice. The cash flow timeline of a paused rainbow annuity immediate is shown on the cover page of this book.

This financial product combines an increasing annuity immediate that makes n payments and a deferred decreasing annuity immediate that also makes n payments. The increasing annuity component is worth $(Ia)_{\overline{n}|i}$ at time 0, while the decreasing annuity component is worth $(Da)_{\overline{n}|i}$ at time n. The latter value should be discounted to time 0 using the discount factor v^n. Thus, the price P^{PRAI} of a paused rainbow annuity immediate is

$$P^{\text{PRAI}} = (Ia)_{\overline{n}|i} + v^n\, (Da)_{\overline{n}|i}. \qquad (201)$$

---

18. Hint:

$$(Ia)_{\overline{n}|i} + v^n\, (Da)_{\overline{n-1}|i} = \frac{\ddot{a}_{\overline{n}|i} - n\, v^n}{i} + v^n\, \frac{n - 1 - a_{\overline{n-1}|i}}{i} = \frac{\ddot{a}_{\overline{n}|i}}{i} + v^n\, \frac{-\ddot{a}_{\overline{n}|i}}{i}, \qquad (198)$$

where we use Eqs (104), (108), and (72). Then, using Eq. (64), we obtain:

$$(Ia)_{\overline{n}|i} + v^n\, (Da)_{\overline{n-1}|i} = \ddot{a}_{\overline{n}|i}\, \frac{1 - v^n}{i} = \ddot{a}_{\overline{n}|i}\, a_{\overline{n}|i}. \qquad (199)$$

Alternatively, we can show that [19]

$$P^{\text{PRAI}} = \ddot{a}_{\overline{n+1}|i}\, a_{\overline{m}|i}. \qquad (205)$$

---

19. Hint:

$$(Ia)_{\overline{m}|i} + v^n\,(Da)_{\overline{m}|i} = \frac{\ddot{a}_{\overline{m}|i} - n\,v^n}{i} + v^n\,\frac{n - a_{\overline{m}|i}}{i} = \frac{\ddot{a}_{\overline{m}|i}}{i} - v^n\,\frac{a_{\overline{m}|i}}{i}, \qquad (202)$$

where we use Eqs (104) and (108). Then, we use Eq. (71) and we compute

$$(Ia)_{\overline{m}|i} + v^n\,(Da)_{\overline{n-1}|i} = a_{\overline{m}|i}\left(\frac{1+i}{i} - \frac{v^n}{i}\right) = a_{\overline{m}|i}\left(1 + \frac{1 - v^n}{i}\right), \qquad (203)$$

so that, using Eqs (64) and (72),

$$(Ia)_{\overline{m}|i} + v^n\,(Da)_{\overline{n-1}|i} = a_{\overline{m}|i}\,(1 + a_{\overline{m}|i}) = a_{\overline{m}|i}\,\ddot{a}_{\overline{n+1}|i}. \qquad (204)$$

Chapter 4

Loans

This chapter is dedicated to loans. By studying them, you learn some financial mathematics, you prepare yourself for an exam, but you also get useful skills in anticipation of the day when you subscribe a loan for buying your first house or apartment. This chapter mainly deals with **amortization**, which consists in reimbursing a loan with a series of - typically - equal payments. From year 2019 on, no more questions are asked on sinking funds. To illustrate this topic, which was considered interesting for a long time by the SOA, I give a very brief overview of these funds, though. The reader only interested in validating the current exam FM can skip this short second part.

4.1 Amortization Method

An individual who enters a loan borrows money for a certain number of years called the **term of the loan**. The borrower reimburses the money he or she borrowed by making several payments, which we assume constant in most situations. The time between any two payment times is called the **payment period** and it is also assumed constant in most situations. When n cash flows are made, and when they are made at the end of every year to reimburse the loan, the term of the loan is also equal to n.

Let i be the effective **interest rate** that we use to compute the value of the regular payments and let X be the total **amount borrowed** at time 0, also called **principal amount**. Then, at the end of every period, the borrower makes a constant **payment amount** P that solves the following equation:

$$X = P\, a_{\overline{n}| i} = P\, \frac{1 - v^n}{i}, \qquad (206)$$

where we simply write that the present value of the amount borrowed should be equal to the present value of all of the cash flows that are paid back. Because equal cash flows are assumed paid back at period ends, the present value of the future cash flows takes the form of a classic annuity immediate.

Note that the constant amount P can be written at any payment time k as follows:

$$P = P_k + I_k, \qquad (207)$$

so as the sum of a **principal repayment** P_k and an **interest payment** I_k. Both the principal repayment and the interest rate payment vary with time, but P does not. Be extremely careful with this notation because the amount P that is paid at the end of every period is not a pure principal payment: it contains both a principal and an interest component. However, P_k is a pure principal component [1].

The **total amount paid**, or sum of the payments made by the borrower during the life of the loan, is

$$A^{\text{tot}} = n\, P, \qquad (208)$$

or, using Eq. (206),

$$A^{\text{tot}} = n\, \frac{X}{a_{\overline{n}| i}}. \qquad (209)$$

The **total amount of interest paid** by the borrower during the life of the loan is simply defined by

$$I^{\text{tot}} = A^{\text{tot}} - X, \qquad (210)$$

---

1. I chose to keep this notation, which may be misleading, to avoid providing inconsistent equations to readers who compare the exercises of the SOA with this book.

so, as the difference between the sum of the reimbursements and the amount borrowed.

Using Eq. (209), we obtain:

$$I^{\text{tot}} = X \left(\frac{n}{a_{\overline{n}|i}} - 1 \right), \qquad (211)$$

which expresses the total amount of interest paid as a function of the amount borrowed.

Alternatively, using Eq. (206), we can write:

$$I^{\text{tot}} = P \left(n - a_{\overline{n}|i} \right), \qquad (212)$$

which expresses the total amount of interest paid as a function of the constant payment.

It is practically and theoretically very important to determine the proportions of principal reimbursement P_k and of interest payment I_k in Eq. (207). As already mentioned, these quantities vary with time. Specifically, I_k is higher at the beginning of the loan and P_k is higher at the end of the loan.

Let k be the period that corresponds to the time interval $[k-1, k)$. For instance, the first period corresponds to the time interval $[0, 1)$, the second period corresponds to the time interval $[1, 2)$, and so on.

To compute I_k and P_k, we need to introduce a bit of additional material. We define the outstanding amount on the loan, or **outstanding balance**, as the **amount that remains due by the borrower to the lender at any point of time**. Clearly, this amount is equal to X at time 0 and to 0 after the term of the loan has elapsed.

In a **prospective approach**, we compute the outstanding balance OB as the Present Value of the Future *principal and interest* Payments, or PVFP [2]. At time k, it is equal to

$$\text{OB}_k = P \, a_{\overline{n-k}|i}, \qquad (213)$$

---

[2]. Although the outstanding balance refers to an amount of *principal* due, do not forget when discounting future cash flows to include *interest rate* cash flows.

because $n-k$ payments remain. Indeed, we can interpret the outstanding balance as 'the current value of what remains to be paid'.

Combining Eqs (206) and (213), we can derive the following useful result:
$$\text{OB}_k = X \, \frac{a_{\overline{n-k}|i}}{a_{\overline{n}|i}}. \tag{214}$$

Also observe that the outstanding balance for period k is the outstanding balance at time $k-1$, or $P \, a_{\overline{n-k+1}|i}$, because $n-k+1$ payments remain in that case.

In the general case where P might not be constant, the **outstanding balance** OB is computed **prospectively** at any time k as follows:
$$\text{OB}_k = \sum_{j=k+1}^{n} \frac{P_j + I_j}{(1+i)^j} = \sum_{j=k+1}^{n} (P_j + I_j) \, v^j. \tag{215}$$

In a **retrospective approach**, the **outstanding balance** OB is the difference between the initial amount of principal due compounded from time 0 to the current time and the past payments also compounded from time 0 to the current time. Therefore, in a retrospective approach we compute the outstanding balance as the difference between 'the current value of what was originally owed' and 'the current value of what has already been paid'. We can write at any time k:
$$\text{OB}_k = X \, (1+i)^k - P \, s_{\overline{k}|i}, \tag{216}$$

when the payments are constant and equal to P.

It can be shown that the prospective and the retrospective approaches of Eqs (213) and (216) produce the same amount OB[3].

---
3. To show that
$$X \, (1+i)^k - P \, s_{\overline{k}|i} = P \, a_{\overline{n-k}|i}, \tag{217}$$
we can prove that
$$v^k \, P \, \left(a_{\overline{n-k}|i} + s_{\overline{k}|i}\right) = X. \tag{218}$$

The amount of interest paid at time k for the kth period is proportional to the outstanding balance at the beginning of the period:
$$I_k = i\ \text{OB}_{k-1}. \tag{222}$$

Therefore,
$$I_k = i\ P\ a_{\overline{n-k+1}|i} = P\left(1 - v^{n-k+1}\right). \tag{223}$$

From the previous result, we can directly compute the amount of principal repaid at time k for the kth period, which is equal to
$$P_k = P - I_k = P\ v^{n-k+1}. \tag{224}$$

A useful alternative expression of P_k is as follows [4]:
$$P_k = \text{OB}_{k-1} - \text{OB}_k. \tag{227}$$

The latter equation confirms that we can interpret the outstanding balance OB_k as an amount of principal owed at time

---

We use the fact that
$$s_{\overline{k}|i} = \sum_{j=0}^{k-1}(1+i)^j = \sum_{j=0}^{k-1}\frac{1}{v^j} = \sum_{j=1-k}^{0} v^j. \tag{219}$$

Then,
$$v^k\ P\left(a_{\overline{n-k}|i} + s_{\overline{k}|i}\right) = v^k\ P\left(\sum_{j=1}^{n-k} v^j + \sum_{j=1-k}^{0} v^j\right) = v^k\ P \sum_{j=1-k}^{n-k} v^j, \tag{220}$$

so that,
$$v^k\ P\left(a_{\overline{n-k}|i} + s_{\overline{k}|i}\right) = P \sum_{j=1}^{n} v^j = P\ a_{\overline{n}|i} = X. \tag{221}$$

4. We can use Eq. (213) to show that
$$\text{OB}_{k-1} - \text{OB}_k = P\left(a_{\overline{n-k+1}|i} - a_{\overline{n-k}|i}\right) = P\left(\frac{1-v^{n-k+1}}{i} - \frac{1-v^{n-k}}{i}\right), \tag{225}$$

so that
$$\text{OB}_{k-1} - \text{OB}_k = P\ \frac{v^{n-k} - v^{n-k+1}}{i} = P\ v^{n-k+1}\ \frac{\frac{1}{v} - 1}{i} = P\ v^{n-k+1}, \tag{226}$$

which is Eq. (224).

Financial Mathematics

k. Indeed, the difference between the amount of principal owed at time $k-1$ and the amount of principal owed at time k is simply the amount of principal that has been reimbursed by the borrower to the lender in the kth period.

Period	OB_{k-1}	I_k	P_k	$I_k + P_k$
1	$X = P\, a_{\overline{m}|i}$	$P\, a_{\overline{m}|i}\, i$	$P - P\, a_{\overline{m}|i}\, i$	P
2	$P\, a_{\overline{n-1}|i}$	$P\, a_{\overline{n-1}|i}\, i$	$P - P\, a_{\overline{n-1}|i}\, i$	P
3	$P\, a_{\overline{n-2}|i}$	$P\, a_{\overline{n-2}|i}\, i$	$P - P\, a_{\overline{n-2}|i}\, i$	P
$\vdots$	$\vdots$	$\vdots$	$\vdots$	$\vdots$
k	$P\, a_{\overline{n-k+1}|i}$	$P\, a_{\overline{n-k+1}|i}\, i$	$P - P\, a_{\overline{n-k+1}|i}\, i$	P
$\vdots$	$\vdots$	$\vdots$	$\vdots$	$\vdots$
$n-1$	$P\, a_{\overline{2}|i}$	$P\, a_{\overline{2}|i}\, i$	$P - P\, a_{\overline{2}|i}\, i$	P
n	$\frac{P}{1+i}$	$\frac{P}{1+i}\, i$	$P - \frac{P}{1+i}\, i = \frac{P}{1+i}$	P

Table 4 – Loan Amortization Table (First Form).

Tables 4 and 5 show the evolutions of OB_{k-1}, I_k, P_k, and $I_k + P_k$, when k ranges between 1 and n. The outstanding balance OB_{k-1}, which is the present value of the remaining payments, naturally decreases with k. The interest payment I_k, which is proportional to the outstanding balance, also decreases with k. The principal repayment, equal to the difference of a constant value and a decreasing function of k, increases with k. The fact that the principal repayment increases geometrically at the constant rate $1 + i$ can be useful for solving some exercises. Finally and by construction, $I_k + P_k$ is always constant. Not written in these tables is the fact that $\text{OB}_n = 0$, which simply expresses that when the term has been reached, nothing remains to be paid.

Quite often, we are given an annual interest rate $i^{(m)}$ convertible m times a year and a maturity equal to n years. In that case, the number of periods is $m \cdot n$ and the interest rate that must

61

Period	OB_{k-1}	I_k	P_k	$I_k + P_k$
1	$X = \frac{P(1-v^n)}{i}$	$P(1-v^n)$	$P v^n$	P
2	$\frac{P(1-v^{n-1})}{i}$	$P(1-v^{n-1})$	$P v^{n-1}$	P
3	$\frac{P(1-v^{n-2})}{i}$	$P(1-v^{n-2})$	$P v^{n-2}$	P
$\vdots$	$\vdots$	$\vdots$	$\vdots$	$\vdots$
k	$\frac{P(1-v^{n-k+1})}{i}$	$P(1-v^{n-k+1})$	$P v^{n-k+1}$	P
$\vdots$	$\vdots$	$\vdots$	$\vdots$	$\vdots$
$n-1$	$\frac{P(1-v^2)}{i}$	$P(1-v^2)$	$P v^2$	P
n	$P v$	$P(1-v)$	$P v$	P

Table 5 – Loan Amortization Table (Second Form).

be used for discounting purposes is $\frac{i^{(m)}}{m}$. Then, Eq. (206), which relates the total amount of the loan X to the regular constant payment P, becomes

$$X = P\, a_{\overline{m \cdot n}|\frac{i^{(m)}}{m}}, \qquad (228)$$

where the related valuation equations should be changed accordingly.

Sometimes, the **last payment** is larger or smaller than the other payments. When the last payment is smaller than P, we call it a **drop payment**; when it is larger than P, we call it a **balloon payment**.

Assume that a last payment equal to Y is made at time $n'+1$ after n' level payments, where $n'+1$ is the term of the loan. In that case, the equation of value can be written as follows:

$$X = P\, a_{\overline{n'}|i} + \frac{Y}{(1+i)^{n'+1}} = P\, \frac{1-v^{n'}}{i} + \frac{Y}{(1+i)^{n'+1}}. \qquad (229)$$

When Y is smaller than P, the situation that is described by Eq. (229) can be made equivalent (in the valuation sense) to the case of a loan with n equal payments, where n is a hypothetical non-integer value such that $n' < n < n' + 1$. When Y is larger than P and smaller than $2P$, the situation of Eq. (229) can be made equivalent to the case of a loan with n equal payments, where n is a hypothetical non-integer value such that $n' + 1 < n < n' + 2$, and so on.

Chapter 5 shows that bonds can be priced by an equation such as Eq. (229).

4.2 Sinking Fund Method

The main concepts associated with a sinking fund are as follows:
- A loan is contracted by a borrower.
- Interest payments are made regularly by the borrower to the lender
- No early principal payments are made to the lender, contrary to what happens with the amortization method.
- Instead, payments are made regularly into a fund called a **sinking fund**.
- At the time of the loan maturity, the amount compounded in the sinking fund serves to repay the principal of the loan to the lender.
- Depending on how the sinking fund is managed, a surplus can exist at the loan maturity if the sinking fund value exceeds the principal of the loan.

Let m be the amount that is paid into a sinking fund and that grows at the rate i_S (this notation is for an effective compound rate, not a simple rate). Assume for instance that payments are made into this sinking fund at the end of every year. The amount M_T that is accumulated at the time T of the reimbursement of the loan is

$$M_T = m \; s_{\overline{m}|i_S}, \qquad (230)$$

where $s_{\overline{m}|i_S}$ is computed using Eq. (66). Let P be the principal amount to be reimbursed on the loan. Then, the amount that remains after repaying the loan is $M_T - P$.

If the sinking fund is not expected to produce any surplus or to show any loss, the level payments m made into the sinking fund should simply satisfy:

$$P = m\ s_{\overline{n}|i_S}. \tag{231}$$

Chapter 5

Bonds

Companies raise funds by borrowing money from banks and by issuing stocks and bonds. The previous chapter dealt with loans and the next chapter will partly deal with stocks. This chapter is dedicated to bonds, which are a key object of study in financial mathematics and, more broadly, in fixed income.

5.1 Definitions and Valuation

When a bond is issued on a primary market, it allows a company to raise an amount equal to - or indexed on - a quantity F called the **par value**, or **face value**.

The **coupon payments** R, which are amounts paid at a regular frequency by a company to its bondholders before and until the maturity of a bond, are proportional to the par value. Indeed, we have:
$$R = F\,r, \qquad (232)$$
where r is the **coupon rate**.

Let n be the **number of coupon payments**. The **term of a bond** is its total period of existence. When payments are annual, the term of the bond is n ; when payments are semiannual, the term of the bond is $n/2$, and so on.

Then, note that the last payment made by a company to its bondholders is not equal to the constant coupon amount. Indeed, when a bond matures, an amount that is either equal to the par value F, or indexed on it, is paid in addition to the final coupon amount.

Because there are many situations where more or less than F is repaid at the maturity, we prefer to call **redemption value** the amount C that is repaid at the maturity in addition to the final coupon amount. Thus, the total payment made at the maturity is $C + R = C + F\ r$. See Fig. 18 for a cash flow timeline.

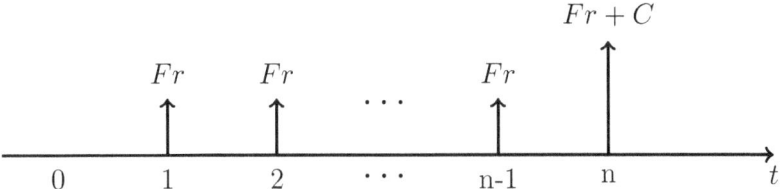

Figure 18 – Cash Flow Timeline of a Bond

Let us now come to the valuation of bonds on secondary markets. The goal of this chapter is not to enter into a sophisticated discussion on the difference between the concepts of value and price, especially because there is no unambiguous definition of these two terms. We simply say that the **price** of a bond is the amount that a counterparty agrees to pay, or could agree to pay, to buy this bond from another counterparty. We also define the **book value** of a bond as the amount assessed by an investor at the time of appraising her or his assets.

The book value is computed as a **Present Value of Future Payments**, or PVFP. Because a variety of effective interest rate values i can be used to compute this present value, several book values $\text{BV}^{[i]}$ exist. Assume that the redemption value C of the bond will be paid in n periods and that n coupons equal to R will be paid at the end of each period. Then, using the effective interest rate i, we obtain:

$$\text{BV}^{[i]} = R\ a_{\overline{n}|i} + C\ v^n = F\ r\ a_{\overline{n}|i} + C\ v^n. \tag{233}$$

On purpose, no time is specified in the above equation. Indeed, you can use Eq. (233) at time 0 and interpret n as the total number of coupon payments made during the life of the bond, or you can use Eq. (233) at any time $t > 0$, where n becomes the number of remaining coupon payments.

Let P be the price of the bond, either quoted or agreed on by two counterparties willing to transact. The annual **effective yield** or **yield rate** j is the solution to

$$P - \left[R\, a_{\overline{n}|j} + C \left(\frac{1}{1+j} \right)^n \right] = 0. \qquad (234)$$

Equivalently, j solves:

$$P = \text{BV}^{[j]}, \qquad (235)$$

so that the yield is the effective interest rate that makes the book value and the price equal.

In the current set of exercises, you do not need to master the distinction between the concepts of price and book value. These two concepts are used interchangeably and the effective interest rate used for valuation purposes is assumed to be identical to the yield rate.

To conclude, we use the following formula for pricing bonds:

$$P = R\, a_{\overline{n}|j} + C \left(\frac{1}{1+j} \right)^n = F\, r\, a_{\overline{n}|j} + C\, v^n. \qquad (236)$$

A very important general result about bonds is as follows:

$$\text{Price} \nearrow \; \Leftrightarrow \; \text{Yield} \searrow, \qquad (237)$$

where you need to keep this in mind when you read the Washington Post, the Financial Times, or browse economic news on the internet. Remember! Bond prices go up when yields go down, and conversely.

The previous relation is valid for someone who considers buying a bond but still does not own it. Indeed, from an exterior viewpoint, all else kept equal, a bond that you buy at a cheaper price provides you with a higher return on your investment, because the coupon payments are fixed [1].

---

1. Reasoning may be more complicated when you already own the bond. In that case, you have to clearly distinguish the price at which you bought the bond and the current price of this bond, were you to sell it on the market.

By analogy with the computation of the present value of a bond, we can derive the future value of a bond as follows:

$$\mathrm{FV}^{[j]} = R\, s_{\overline{n}|j} + C = F\, r\, s_{\overline{n}|j} + C, \qquad (238)$$

where the valuation is performed at the time of the last coupon payment and the prevailing yield is j.

In the case of bonds with semiannual coupons, the number n of payments is of course equal to TWICE the term of the bond expressed in years (do not forget this factor 2, which is introduced in many exercises!) and the semiannual yield rate j can be computed as follows:

$$j = \frac{i^{(2)}}{2}, \qquad (239)$$

when we are given a nominal annual interest rate $i^{(2)}$ convertible semiannually.

For such bonds, we are often given a **nominal annual coupon rate convertible semiannually** e equal to $2\,r$ (be careful of this factor 2!) and a par value F. Therefore, the **semiannual coupon rate** is equal to

$$r = \frac{e}{2}, \qquad (240)$$

while a payment of

$$R = F\, r \qquad (241)$$

is made every six months on this bond.

Similar proportionality operations are performed in the case of bonds with quarterly payments, monthly payments, and so on.

We say that a bond is **redeemable at par** when

$$C = F, \qquad (242)$$

so when the redemption value is equal to the par value.

Also note that the **total interest** paid on a standard bond is equal to

$$I = n\, r\, F, \qquad (243)$$

where n is the number of coupons, F is the face value, and r is equal to the coupon rate in the case of **annual payments**, to *half* the annual nominal coupon rate in the case of **semiannual payments**, and so on.

There are some situations where we do not use a unique effective interest rate for pricing purposes, but instead several **spot** interest rates s_k.

A spot interest rate is an interest rate that prevails at the valuation time and that covers a specific period. Assume that the valuation time is 0. Then, each spot interest rate s_k covers the period $(0, k)$.

In such a situation, the price of a bond becomes

$$P = \sum_{k=1}^{n} \frac{R_k}{(1+s_k)^k} + \frac{C}{(1+s_n)^n}, \qquad (244)$$

where the knowledge of $(s_1, s_2, \cdots, s_n)$ is required to price this bond.

5.2 Zero-Coupon Bonds

The short term debt of companies and governments often takes the form of **zero-coupon bonds**. By definition, these bonds do not pay coupons. Instead, they reimburse at their maturity an amount that is superior to the amount borrowed at time 0.

By convention, the amount reimbursed at the maturity is often set equal to 1, while the amount initially borrowed can be 0.98, 0.99, or any similar amount.

Let $Z(0, n)$ be the price at time 0 of a zero-coupon bond that promises one currency unit at time n. We write:

$$Z(0,n) < 1 \quad \text{at } t = 0 \quad \to \quad 1 \quad \text{at } t = n. \qquad (245)$$

Conversely, one unit of cash invested in a zero-coupon bond produces $1/Z(0,T)$ units of cash as time n. Then, we have:

$$1 \quad \text{at } t = 0 \quad \to \quad \frac{1}{Z(0,n)} > 1 \quad \text{at } t = n. \qquad (246)$$

The market price $Z(0, n)$ of a zero-coupon bond is related to its yield rate j as follows:

$$Z(0, n) = \frac{1}{(1+j)^n}, \qquad (247)$$

where this pricing formula can be recovered from Eq. (236) by setting $C = 1$ and $R = 0$.

5.3 Premium and Discount Bonds

An example of a bond whose price is F, also called a bond that **sells at par** or a **par value bond**, is illustrated by the following pricing equation [2]:

$$F = F\, j\, a_{\overline{n}|j} + F\left(\frac{1}{1+j}\right)^n = F\, j\, a_{\overline{n}|j} + F\, v^n, \qquad (248)$$

which holds when the coupon rate r is equal to the yield j and when the redemption value C is equal to the face value F.

Of course, Eq. (248) is equivalent to

$$C = C\, j\, a_{\overline{n}|j} + C\left(\frac{1}{1+j}\right)^n = C\, j\, a_{\overline{n}|j} + C\, v^n. \qquad (249)$$

Using Eq. (236), we can derive the following properties for a bond that is redeemable at par $(C = F)$:
— If the yield is equal to the coupon rate $(j = r)$, then the price is equal to the face value $(P = \mathrm{BV}^{[j]} = C = F)$.
— If the price is equal to the face value $(P = \mathrm{BV}^{[j]} = C = F)$, then the yield is equal to the coupon rate $(j = r)$.

Bonds redeemable at par are sometimes improperly called par value bonds. As can be seen from Eq. (248), for bonds redeemable at par to be par value bonds, the additional constraint $j = r$ should be satisfied.

This section dedicated to premium and discount bonds will study in a systematic way the real-world situations where bond prices are not equal to C.

---

2. To derive this equation, use the expression of $a_{\overline{n}|j}$ provided in Eq. (64).

5.3.1 Main Definitions

From now on and for simplicity, we drop the superscript $[j]$ in the notation of BV and we assume that the book value is computed with the yield so that it is equal to the price. Further, we write that $BV(t)$ is the book value of a bond at time t.

A bond that **sells at a premium**, or **premium bond**, is a bond that satisfies
$$BV(0) > C, \qquad (250)$$
where the **premium** is the difference $BV(0) - C$.

If a bond sells at a premium, the value of its premium can be derived by computing the difference of Eqs (236) and (249):
$$BV(0) - C = (F\,r - C\,j)\,a_{\overline{n}|j}. \qquad (251)$$

We can decompose Eq. (251) cash flow by cash flow using Eq. (63). The **present value of the premium due to the kth cash flow** is
$$(F\,r - C\,j)\,v^k. \qquad (252)$$

Then, we define the **modified coupon rate** as follows:
$$r^C = \frac{R}{C} = \frac{r\,F}{C}. \qquad (253)$$

From Eq. (251), we see that
$$BV(0) > C \quad \Leftrightarrow \quad F\,r > C\,j. \qquad (254)$$

From Eqs (253) and (254), we deduce that a **premium bond** is a bond that satisfies
$$r^C > j, \qquad (255)$$
which is a comparison in terms of rates, while Eq. (250) was a comparison in terms of prices.

For a bond redeemable at par ($C = F$), the condition shown in Eq. (255) is equivalent to
$$r > j, \qquad (256)$$

while the condition shown in Eq. (250) is equivalent to

$$BV(0) > F. \qquad (257)$$

A bond **sells at a discount**, or **discount bond**, is a bond that satisfies

$$BV(0) < C, \qquad (258)$$

where the **discount** is the difference $C - BV(0)$.

If a bond sells at a discount, the value of its discount can be derived by computing the difference of Eqs (249) and (236):

$$C - BV(0) = (C\ j - F\ r)\ a_{\overline{n}|j}. \qquad (259)$$

We can decompose Eq. (259) cash flow by cash flow using Eq. (63). The **present value of the discount due to the kth cash flow** is

$$(C\ j - F\ r)\ v^k. \qquad (260)$$

From Eq. (259), we see that

$$C > BV(0) \quad \Leftrightarrow \quad C\ j > F\ r. \qquad (261)$$

From Eqs (253) and (261), we deduce that a discount bond is a bond that satisfies

$$r^C < j. \qquad (262)$$

For a bond redeemable at par $(C = F)$, the condition shown in Eq. (262) is equivalent to

$$r < j, \qquad (263)$$

while the condition shown in Eq. (258) is equivalent to

$$BV(0) < F. \qquad (264)$$

5.3.2 Book Value Evolution

Based on Eq. (236), we write at any time k, **immediately after** the payment of the kth coupon:

$$BV(k) = F\ r\ a_{\overline{n-k}|j} + C\ v^{n-k}. \qquad (265)$$

Next, we use the following generalization of Eq. (251) at any time k:
$$\text{BV}(k) - C = (F\ r - C\ j)\ a_{\overline{n-k}|j}. \qquad (266)$$

We compute:
$$\text{BV}(k) - \text{BV}(k-1) = (F\ r - C\ j)\left[a_{\overline{n-k}|j} - a_{\overline{n-k+1}|j}\right], \qquad (267)$$

so that
$$\text{BV}(k) - \text{BV}(k-1) = -(F\ r - C\ j)\ v^{n-k+1}, \qquad (268)$$

using the definition of $a_{\overline{n-k}|j}$ in Eq. (63).

For a discount bond, we compute the **amount for accumulation of discount in the kth coupon** as follows:
$$\text{BV}(k) - \text{BV}(k-1) = (C\ j - F\ r)\ v^{n-k+1}, \qquad (269)$$

because in the case of a discount, $\text{BV}(k)$ progressively **increases** with time, so $\text{BV}(k) - \text{BV}(k-1) > 0$ and there is a **write-up** of the discount.

Similarly, for a premium bond, we compute the **amount for amortization of premium in the kth coupon** as follows:
$$\text{BV}(k-1) - \text{BV}(k) = (F\ r - C\ j)\ v^{n-k+1}, \qquad (270)$$

because in the case of a premium, $\text{BV}(k)$ progressively **decreases** with time, so $\text{BV}(k) - \text{BV}(k-1) < 0$, and there is a **write-down** of the premium.

5.3.3 Comparison of Bonds and Loans

By analogy with Eq. (227) that identifies the difference between the outstanding balances of a loan at times $k-1$ and k to the amount of principal that is reimbursed at time k, we can identify $\text{BV}(k-1) - \text{BV}(k)$ to the amount of principal P_k that is reimbursed in the kth coupon:
$$P_k = \text{BV}(k-1) - \text{BV}(k) = (F\ r - C\ j)\ v^{n-k+1}, \qquad (271)$$

where you should be careful that this quantity can be positive of negative, contrary to what happens with a standard loan.

The amount of interest I_k that is paid in the kth coupon is simply equal to the difference between the value of this coupon and the amount of principal reimbursed in this coupon. Therefore, we have:

$$I_k = F\,r - P_k = F\,r\,(1 - v^{n-k+1}) + C\,j\,v^{n-k+1}. \qquad (272)$$

We can recover Eq. (272) by computing

$$I_k = j\,\text{BV}(k-1), \qquad (273)$$

using Eq. (265).

By analogy with Tables 4 and 5 that show the amortization structure of a loan, we present in Table 6 the amortization structure of a bond. In the last row of this table we indicate by $(+C)$ the possibility to include the redemption amount in the definition of $I_n + P_n$. In that case, P_n should be inflated accordingly.

Note finally that P_k is positive in the case of a premium bond and can be identified to the amount for amortization of premium of Eq. (270). An investor who buys a premium bond pays by definition a high buying price. Positive principal repayments are made to him or her already in the coupons. Therefore, the amount of interest paid in each coupon is inferior to the coupon amount itself. We summarize that by saying that for a premium bond, $P_k > 0$ and $I_k < F\,r$.

Everything is reversed in the case of a discount bond, where P_k is negative and can be identified to *minus* the amount for accumulation of discount of Eq. (269). An investor who buys a discount bond pays by definition a low buying price. She should compensate for this low price by making additional payments of principal during the life of the bond. These additional payments impact the coupon payments. Indeed, we can consider that negative principal payments are made by the issuer to the investor in the coupons (so that the investor makes positive principal payments to the issuer). Therefore, the amount of interest paid in each coupon is superior to the coupon amount itself. We

Period	BV_{k-1}	I_k	P_k	$I_k + P_k$
1	$F\,r\,a_{\overline{n}|j} + C\,v^n$	$F\,r\,(1-v^n) + C\,j\,v^n$	$(F\,r - C\,j)\,v^n$	$F\,r$
2	$F\,r\,a_{\overline{n-1}|j} + C\,v^{n-1}$	$F\,r\,(1-v^{n-1}) + C\,j\,v^{n-1}$	$(F\,r - C\,j)\,v^{n-1}$	$F\,r$
3	$F\,r\,a_{\overline{n-2}|j} + C\,v^{n-2}$	$F\,r\,(1-v^{n-2}) + C\,j\,v^{n-2}$	$(F\,r - C\,j)\,v^{n-2}$	$F\,r$
...	...	...	...	...
k	$F\,r\,a_{\overline{n-k+1}|j} + C\,v^{n-k+1}$	$F\,r\,(1-v^{n-k+1}) + C\,j\,v^{n-k+1}$	$(F\,r - C\,j)\,v^{n-k+1}$	$F\,r$
...	...	...	...	...
$n-1$	$F\,r\,a_{\overline{2}|j} + C\,v^2$	$F\,r\,(1-v^2) + C\,j\,v^2$	$(F\,r - C\,j)\,v^2$	$F\,r$
n	$(F\,r + C)\,v$	$F\,r\,(1-v) + C\,j\,v$	$(F\,r - C\,j)\,v\,(+C)$	$F\,r\,(+C)$

Table 6 – Bond Amortization Table.

summarize that by saying that for a discount bond, $P_k < 0$ and $I_k > F\ r$.

5.3.4 Final Remark

Observe that $\text{BV}(k) \cdot (1+j)$ is the book value of the bond compounded from the time that immediately follows the kth coupon payment to the time that immediately precedes the $(k+1)^{\text{th}}$ coupon payment. Put more simply, $\text{BV}(k) \cdot (1+j)$ is the book value of the bond **immediately before** the $(k+1)^{\text{th}}$ coupon payment.

However, $\text{BV}(k+1)$ is the book value of the bond **immediately after** the $(k+1)^{\text{th}}$ coupon payment. Therefore, we readily have:
$$\text{BV}(k) \cdot (1+j) - \text{BV}(k+1) = F\ r, \tag{274}$$
which expresses that the difference in book values before and after a coupon payment is simply the amount of this coupon payment. Clearly, this relation holds for any value of k.

5.4 Callable Bonds

A **callable bond** is a bond that can be redeemed in advance by the issuer. The possibility for the company to reimburse this bond early is a right, or **option**. You will learn about options and other derivatives when you study for the exam IFM of the SOA. A **non-callable** bond is a bond for which the issuer does not own such an option.

When a bond is early reimbursed, or called, or again withdrawn, an amount that can be different from C or F is paid. Let W be the amount that is reimbursed if the bond is called. After that, no more payments are made.

The **modified coupon rate** is now defined as follows:
$$r^W = \frac{R}{W} = \frac{r\ F}{W}, \tag{275}$$
where the difference with Eq. (253) is that the denominator becomes equal to W.

Consistently with the exercises, we also define a **lowest semi-annual yield rate** j by

$$j = \frac{i^{(2)}}{2}, \qquad (276)$$

where we are given a **lowest nominal annual interest rate** $i^{(2)}$ convertible semiannually.

We assume that the bond can be called on a discrete set of dates. Outside these dates, the company does not have the option to force early redemption. Thus, we assume that there are M call dates and we denote these call dates by $t_{k=1,\ldots,M}$.

Then, two main situations arise. The first of these situations can be summarized as follows:

$$r^W > j \Rightarrow \text{Premium Bond} \Rightarrow WCIER \Rightarrow n_W = \min_{k=1,\ldots,M} t_k, \qquad (277)$$

where n_W is the time of early redemption, WCIER stands for "Worst Case If Early Redemption", and this worst case is for the bondholder.

This series of implications expresses that when the modified coupon rate is too high compared to the yield, the issuer is paying too generous coupons to bondholders. Therefore, the issuer had better early reimburse as soon as possible in order to refinance itself on the market at the rate of j. The fact that the modified coupon rate is too high corresponds to a bond that is too expensive, so that sells at a premium.

Another way to look at that is to say that issuers prefer bonds with a lower value, which means that they owe less to their lenders. Thus, issuers will try to lower the value of their bonds (equivalently to offer the smallest possible *actual* yield) and to reach a situation that is most detrimental to lenders. Here, early redemption is most advantageous to the issuer and most disadvantageous to the bondholder. Therefore, it will occur as soon as possible, at time $n_W = \min_{k=1,\ldots,M} t_k$.

The second classic situation is constructed by inverting all the inequalities and conclusions of the first situation. This second

situation can be summarized as follows:
$$r^W < j \Rightarrow \text{Discount Bond} \Rightarrow WCILR \Rightarrow n_W = \max_{k=1,\ldots,M} t_k, \tag{278}$$
where WCILR stands for "Worst Case If Late Redemption" and this worst case is again for the bondholder.

This series of implications expresses that when the modified coupon rate is too low compared to the yield, the issuer is paying too small coupons to bondholders, compared to market standards. Therefore, the issuer had better reimburse as late as possible. The fact that the modified coupon rate is too low corresponds to a bond that is cheap, so that sells at a discount.

As already mentioned, issuers try to lower the value of their bonds (equivalently to offer the smallest possible *actual* yield) and to reach a situation that is most detrimental to lenders. Here, late redemption is most advantageous to the issuer and most disadvantageous to the bondholder. Therefore, it will occur as late as possible, at time $n_W = \max_{k=1,\ldots,M} t_k$.

In the presence of several call periods and early redemption values (also called withdrawal values), the idea is to compute the smallest price for each of the call periods and to retain the smallest of these smallest prices. As already written, an issuer calls a bond to minimize its value, which is equivalent to minimizing the value of future payments.

Whatever the situation considered, after the optimal (from the viewpoint of the issuer) early redemption time has been determined, it is possible to deduce the actual effective yield offered by the bond.

Based on Eq. (236), we construct the pricing formula of callable bonds as follows:
$$P = R\ a_{\overline{n_W}|j} + W\left(\frac{1}{1+j}\right)^{n_W} = F\ r\ a_{\overline{n_W}|j} + W\ v^{n_W}, \tag{279}$$
where n_W is the **number of payments until the call date**.

When P has already been paid, we can deduce the actual effective yield of the bond using Eq. (279). Conversely, for a target yield j, we can deduce a price P.

Note that when $r^W = j$, $Fr = jW$ and the price of the bond is simply the withdrawal value because

$$jW\, a_{\overline{n_W}|j} + W\, v^{n_W} = W. \qquad (280)$$

From the comparison of Eqs (279) and (280), we see that the price of a premium bond (for which $r^W > j$, or equivalently $R > jW$) is superior to W. Conversely, the price of a discount bond (for which $r^W < j$, or equivalently $R < jW$) is inferior to W.

Chapter 6

General Cash Flows and Portfolios

This chapter presents many useful results on bonds and other financial products. First, we measure the performance of a portfolio when money is deposited into it or withdrawn from it. Then, we show how to compute Macaulay and modified durations and convexities. We illustrate the computation of these indicators on bonds, level annuities and perpetuities, and increasing annuities and perpetuities. Next, we examine the pricing of stocks, before providing approximations for the recomputation of bond prices when yields change. We conclude this chapter by a study of forward interest rates.

6.1 Rates of Return

Imagine that you constitute a fund in which you invest 1 at time 0. The next day, at time $t = \frac{1}{252}$, you make an additional investment of 100 in this fund[1]. The market is supposed not to have moved between times 0 and 1/252. Then, you stop making payments into this fund and you also do not receive payments

---

1. Investments are only assumed possible during open days. There are about 252 open days per year. One day is therefore equal to about 1/252 year.

from this fund until the end of the year. Finally, at time $t = 1$, the total investment is worth 50 due to market fluctuations between times $1/252$ and 1.

What return did you achieve? If you compute the return between times 0 and 1, you conclude that you achieved a great return of $\frac{50-1}{1} = 4,900\%$. However, if you compute the return between times $\frac{1}{252}$ and 1, you conclude that you achieved a catastrophic return of $\frac{50-101}{101} \approx -50.49\%$.

What is the correct conclusion? You can guess that it is probably more the second one, but at the cost of forgetting the information at time 0.

More generally, you understand with this simple example that it can be hard to compute investment returns when inflows of cash are made in the middle of a period under study. The same issue arises when withdrawals of cash occur.

We now introduce two ways of measuring a **yield rate** or **rate of return** when inflows or outflows of cash can be made between the initial and final times of a given investment period.

Let M_0 and M_T be the initial and final values of an investment or account. Deposits of D_{t_j} are made at times t_j and withdrawals of W_{t_k} are made at times t_k.

We define the **dollar-weighted rate of return**, or **money-weighted rate of return**, as follows:

$$\text{DWRR} = \frac{M_T - M_0 - \sum_j D_{t_j} + \sum_k W_{t_k}}{M_0 + \sum_j D_{t_j} \frac{T-t_j}{T} - \sum_k W_{t_k} \frac{T-t_k}{T}}, \qquad (281)$$

where the numerator is the **interest earned during the period** and the denominator is the **average amount exposed to earning interest**.

For illustration, the dollar-weighted rate of return of the investment considered at the beginning of this section is equal to

$$\text{DWRR} = \frac{50 - 1 - 100}{1 + 100 \cdot \frac{1 - \frac{1}{252}}{1}} \approx -50.69\%, \qquad (282)$$

which is quite close to (but different than) the return that was computed using a naive approach between times $\frac{1}{252}$ and 1.

Another way of measuring a rate of return is with the **time-weighted rate of return**, which is defined by

$$\text{TWRR} = \left(\prod_k \frac{M_{t_k^-}}{M_{t_{k-1}}} \right) - 1, \tag{283}$$

where $M_{t_k^-}$ is the amount in the account just before time k and M_{t_k} is the amount in the account at time k immediately after a deposit D_{t_k} and/or a withdrawal W_{t_k} have been made. The amounts $M_{t_k^-}$ and M_{t_k} are related as follows:

$$M_{t_k} = M_{t_k^-} + D_{t_k} - W_{t_k}. \tag{284}$$

In the illustration at the beginning of the section, we had $M_{\left(\frac{1}{252}\right)^-} = 1$, because we assumed that the investment is stable between times 0 and $\frac{1}{252}$. We also had $M_{\frac{1}{252}} = 1 + 100 = 101$. Therefore, we can compute:

$$\text{TWRR} = \frac{M_{\left(\frac{1}{252}\right)^-}}{M_0} \cdot \frac{M_1}{M_{\frac{1}{252}}} - 1 = \frac{1}{1} \cdot \frac{50}{101} - 1 \approx -50.49\%, \tag{285}$$

which is exactly the result of the second naive approach. In this example, TWRR does not allow us to reach a more precise answer than the second naive approach [2].

Consider now another situation where you invest 1 at time 0 in a fund managed by someone else. This amount is inflated at a rate of 4% per year over a period of six months and becomes equal to $1 \cdot 1.04^{\frac{1}{2}}$ at that time. At time $\frac{1}{2}$, you invest 100 in the fund, which becomes therefore equal to $100 + 1.04^{\frac{1}{2}}$. During a second period of six months, the fund manager is able to grow the fund at a rate of 6%. Therefore, you own at time 1 an amount of $\left(100 + 1.04^{\frac{1}{2}}\right) \cdot 1.06^{\frac{1}{2}}$.

---

2. In more realistic examples, the value of the investment varies between times 0 and $\frac{1}{252}$, yielding a value of $M_{\left(\frac{1}{252}\right)^-}$ different from 1 and changing the conclusions. In that case, TWRR is different from the yield computed with the second naive approach.

What should be the rate of return of your investment? Because the manager of the fund in which you invested your money produced a return of 4% during the first six months and a return of 6% during the second six months, your guess is that a fair assessment of the rate of return during the whole year should be something like 5%, independently of the size of your inflows of cash in the fund. Is TWRR able to produce such a result? We compute:

$$\text{TWRR} = \frac{1.04^{\frac{1}{2}}}{1} \cdot \frac{\left(100 + 1.04^{\frac{1}{2}}\right) \cdot 1.06^{\frac{1}{2}}}{100 + 1.04^{\frac{1}{2}}} - 1 = 1.04^{\frac{1}{2}} \cdot 1.06^{\frac{1}{2}} - 1, \qquad (286)$$

so that indeed

$$\text{TWRR} \approx 4.995\% \qquad (287)$$

is close to 5%.

To conclude, DWRR is more appropriate to measure the aggregate performance of *your money* and takes into account the size of your inflows or outflows of cash. On the contrary, TWRR is more appropriate to measure how *fund managers* perform over several sub-periods, independently of how much you invest in their fund.

Finally observe that, hopefully, **in the absence of deposits and withdrawals**, the dollar-weighted rate of return and the time-weighted rate of return satisfy

$$\text{DWRR} = \text{TWRR} = \frac{M_T - M_0}{M_0}, \qquad (288)$$

so are equivalent to an effective rate of interest.

6.2 Macaulay Duration and Convexity

Consider a bond that makes a payment of 5 at time 1 and a payment of 105 at time 2. This bond matures at time 2 but this does not tell us at what **average time** it reimburses a lender. This average time presumably occurs between times 1 and 2. However, if we want to compute it, we need a specific tool.

In 1938, Macaulay introduced such a tool, which is now known as the **Macaulay duration**. This indicator is defined as follows:

$$D^{\text{Mac}} = \frac{\sum_{k=1}^{n} t_k \, CF_{t_k} \, v^{t_k}}{\sum_{k=1}^{n} CF_{t_k} \, v^{t_k}} = \frac{\sum_{k=1}^{n} t_k \, CF_{t_k} \, e^{-\delta t_k}}{\sum_{k=1}^{n} CF_{t_k} \, e^{-\delta t_k}}, \qquad (289)$$

where we compute the arithmetic average of the payment times t_k, using weights equal to the discounted cash flows $CF_{t_k} \, v^{t_k}$.

You can check [3] that the Macaulay duration of a zero-coupon bond is simply equal to its term T.

Observe that the denominator of the Macaulay duration formula, being the sum of discounted cash flows, is also equal to the price P of the product. We denote:

$$P = \sum_{k=1}^{n} CF_{t_k} \, v^{t_k} = \sum_{k=1}^{n} CF_{t_k} \, e^{-\delta t_k}. \qquad (290)$$

By observing that the numerator of Eq. (289) is equal to *minus* the derivative of the price P with respect to the force of interest δ, we are able to rewrite the Macaulay duration formula as follows:

$$D^{\text{Mac}} = -\frac{1}{P} \frac{\partial P}{\partial \delta}. \qquad (291)$$

The Macaulay duration is linear in the sense that the Macaulay duration of a portfolio of securities is a linear combination of the Macaulay durations of the portfolio components, where the weights of the linear combination are the weights of the component securities within the total portfolio.

For instance, let it be a portfolio that consists of asset 1 with weight w_1 and Macaulay duration D_1^{Mac}, and of asset 2 with weight w_2 and Macaulay duration D_2^{Mac}. Then, the Macaulay

---

3. A unique cash flow occurs at time T and the weight at the numerator and at the denominator of the Macaulay duration formula can be simplified.

duration D^{Mac} of the aggregate portfolio is [4]:

$$D^{\text{Mac}} = w_1\, D_1^{\text{Mac}} + w_2\, D_2^{\text{Mac}}, \qquad (294)$$

where a similar result holds when we have more than two sub-portfolios.

Similar to our computation of an average payment time with respect to discounted cash flows, we can compute the average square payment time of a bond with respect to discounted cash flows. This is the **Macaulay convexity**, which is defined as follows:

$$C^{\text{Mac}} = \frac{\sum_{k=1}^{n} t_k^2\, CF_{t_k}\, v^{t_k}}{\sum_{k=1}^{n} CF_{t_k}\, v^{t_k}} = \frac{\sum_{k=1}^{n} t_k^2\, CF_{t_k}\, e^{-\delta t_k}}{\sum_{k=1}^{n} CF_{t_k}\, e^{-\delta t_k}}. \qquad (295)$$

This indicator is proportional to the second order derivative of the price with respect to the force of interest:

$$C^{\text{Mac}} = \frac{1}{P}\, \frac{\partial^2 P}{\partial \delta^2}. \qquad (296)$$

Finally, remember that the Macaulay convexity of a zero-coupon bond is simply equal to the square of this bond maturity, so to T^2. This result is a direct consequence of Eq. (295).

6.3 Modified Duration and Convexity

In 1939, Hicks studied how bond prices evolve when market yields change. This led to a new definition of the concept of

---

4. This can be proved as follows:

$$D^{\text{Mac}} = \frac{\sum_{k=1}^{n} t_k\, CF_{t_k}\, v^{t_k}}{P} = \frac{\sum_{k=1}^{n} t_k\, CF_{t_k}^1\, v^{t_k} + \sum_{k=1}^{n} t_k\, CF_{t_k}^2\, v^{t_k}}{P}, \qquad (292)$$

where $CF_{t_k}^i$ is a cash flow at time t_k of sub-portfolio i. Next, we write:

$$D^{\text{Mac}} = \frac{P_1}{P}\, \frac{\sum_{k=1}^{n} t_k\, CF_{t_k}^1\, v^{t_k}}{P_1} + \frac{P_2}{P}\, \frac{\sum_{k=1}^{n} t_k\, CF_{t_k}^2\, v^{t_k}}{P_2}, \qquad (293)$$

which is our result.

duration. We first define the relative sensitivity of prices with respect to yields, or equivalently effective interest rates, by

$$S = \frac{1}{P}\frac{\partial P}{\partial i}. \tag{297}$$

This is a relative sensitivity indicator because it tells you how a relative change in price $\frac{\partial P}{P}$ is produced when a change in rate ∂i occurs [5].

Then, observe that S is negative because, as already mentioned, prices increase when rates decrease, and conversely. To obtain a positive indicator, we define the modified duration as *minus* the sensitivity. We have:

$$D^{\text{Mod}} = -S = -\frac{1}{P}\frac{\partial P}{\partial i}, \tag{298}$$

where this indicator is called a modified duration because it is not equal to the Macaulay duration.

If you compare Eqs (291) and (298), you observe that they are very similar. You should not forget that in the first of these equations, differentiation is conducted with respect to the force of interest, while in the second of these equations, differentiation is conducted with respect to the effective interest rate (or yield).

Using Eq. (290) and the fact that $\frac{\partial v^{t_k}}{\partial i} = -t_k\, v^{t_k+1}$, we obtain:

$$D^{\text{Mod}} = \frac{\sum_{k=1}^{n} t_k\, CF_{t_k}\, v^{t_k+1}}{\sum_{k=1}^{n} CF_{t_k}\, v^{t_k}}, \tag{299}$$

which enables us to link the Macaulay and modified durations in a simple way:

$$D^{\text{Mod}} = D^{\text{Mac}} \cdot v = \frac{D^{\text{Mac}}}{1+i}. \tag{300}$$

From this result, we conclude that the Macaulay duration can also be used as a measure of interest rate sensitivity, up to a

---

[5]. Of course, this is not at all clean from a mathematical viewpoint because ∂i is an infinitesimal change in rate. The goal here is just to get an intuition of what we are doing.

simple proportionality factor. Indeed, observe that

$$D^{\text{Mac}} = D^{\text{Mod}}\left(1+i\right) = -\frac{1}{P}\frac{\partial P}{\partial i}\left(1+i\right) = -S\left(1+i\right). \quad (301)$$

Finally, remember that the modified duration of a zero-coupon bond is equal to $T\,v$.

A first order measure such as S or D^{Mod} is only meaningful in situations of small changes in interest rates. These measures, similar to prices, also move with interest rates. After a big change in yield, the values of S or D^{Mod} need to be recomputed. An indicator that measures the sensitivity of a sensitivity measure is the convexity. We define the **modified convexity** as follows:

$$C^{\text{Mod}} = \frac{1}{P}\frac{\partial^2 P}{\partial i^2}. \quad (302)$$

Using Eq. (290) and the fact that $\frac{\partial^2 v^{t_k}}{\partial i^2} = t_k\,(t_k+1)\,v^{t_k+2}$, we obtain:

$$C^{\text{Mod}} = \frac{\sum\limits_{k=1}^{n} t_k\,(t_k+1)\,CF_{t_k}\,v^{t_k+2}}{\sum\limits_{k=1}^{n} CF_{t_k}\,v^{t_k}}, \quad (303)$$

which allows us to extend Eq. (300) to the second order as follows:

$$C^{\text{Mod}} = \left(C^{\text{Mac}} + D^{\text{Mac}}\right)\cdot v^2. \quad (304)$$

Finally, note that the modified convexity of a zero-coupon bond is equal to $T\,(T+1)\,v^2$.

6.4 Bonds

We now illustrate the previous two sections with bonds redeemable at par (for which $C = F$). We assume that $t_k = k$ for all k, that the intermediate cash flows are equal to $CF_{t_k} = rF$ for all $k < n$, and that the final cash flow is equal to $CF_{t_n} = F + rF$.

Using Eq. (289), we can show that the **Macaulay duration of such a bond** is equal to

$$D^{\text{Mac}}_{\text{Bond}} = \frac{\sum_{k=1}^{n} k \, r \, F \, v^k + n \, F \, v^n}{\sum_{k=1}^{n} r \, F \, v^k + F \, v^n} = \frac{r \, (Ia)_{\overline{n}|i} + n \, v^n}{r \, a_{\overline{n}|i} + v^n}. \qquad (305)$$

Assume that the bond is priced as par, so that $r = i$. In that situation, we have:

$$D^{\text{Mac}}_{\text{Bond}} = \frac{r \, (Ia)_{\overline{n}|r} + n \, v_r^n}{r \, a_{\overline{n}|r} + v_r^n} = \ddot{a}_{\overline{n}|r}, \qquad (306)$$

where $v_r = \frac{1}{1+r}$. In order to compute the second equality, we simplified the numerator using Eq. (104) and the denominator using Eq. (64).

In the case of a bond that pays semiannual coupons, you are given an annual coupon rate $2 \, r$ paid semiannually. The coupon payments are equal to $F \, r$. If the bond makes n payments, so if it has a maturity of $\frac{n}{2}$, we can derive:

$$D^{\text{Mac}}_{\text{Bond}} = \frac{\sum_{k=1}^{n} \frac{k}{2} \, r \, F \, v_j^k + \frac{n}{2} \, F \, v_j^n}{\sum_{k=1}^{n} r \, F \, v_j^k + F \, v_j^n} = \frac{1}{2} \frac{r \, (Ia)_{\overline{n}|j} + n \, v_j^n}{r \, a_{\overline{n}|j} + v_j^n}, \qquad (307)$$

where j is the effective interest rate over a period of six months, $v_j = \frac{1}{1+j}$, and the payment times are given by $\frac{k}{2}$.

Assume now that the bond is priced at par, so that $j = r$. We obtain:

$$D^{\text{Mac}}_{\text{Bond}} = \frac{1}{2} \frac{r \, (Ia)_{\overline{n}|r} + n \, v_r^n}{r \, a_{\overline{n}|r} + v_r^n} = \frac{\ddot{a}_{\overline{n}|r}}{2}, \qquad (308)$$

where again $v_r = \frac{1}{1+r}$.

Note that for each of the Eqs (305) to (308), we can compute the corresponding modified duration value $D^{\text{Mod}}_{\text{Bond}}$ using Eq. (300).

When the bond is not redeemable at par, so when $C \neq F$, we cannot obtain simple duration formulas such as Eqs (306) and (308). However, using Eq. (289), we can write in the case of annual coupons:

$$D^{\text{Mac}}_{\text{Bond}'} = \frac{\sum_{k=1}^{n} k\, r\, F\, v^k + n\, C\, v^n}{\sum_{k=1}^{n} r\, F\, v^k + C\, v^n} = \frac{r\, F\, (Ia)_{\overline{n}|i} + n\, C\, v^n}{r\, F\, a_{\overline{n}|i} + C\, v^n},$$

(309)

while we have in the case of n semiannual coupons:

$$D^{\text{Mac}}_{\text{Bond}'} = \frac{\sum_{k=1}^{n} \frac{k}{2}\, r\, F\, v_j^k + \frac{n}{2}\, C\, v_j^n}{\sum_{k=1}^{n} r\, F\, v_j^k + C\, v_j^n} = \frac{1}{2} \frac{r\, F\, (Ia)_{\overline{n}|j} + n\, C\, v_j^n}{r\, F\, a_{\overline{n}|j} + C\, v_j^n},$$

(310)

where j is the semiannual effective interest rate.

6.5 Level Annuities and Perpetuities

Let us now consider the case of level annuities, for which all payments are identical, contrary to coupon bonds. We assume that $CF_{t_k} = 1$ and $t_k = k$ for all k. We compute the **Macaulay duration of an annuity immediate** as follows:

$$D^{\text{Mac}}_{\text{AnnImm}} = \frac{\sum_{k=1}^{n} k\, v^k}{\sum_{k=1}^{n} v^k} = \frac{(Ia)_{\overline{n}|i}}{a_{\overline{n}|i}} = \frac{1+i}{i} - \frac{n\, v^n}{1 - v^n}, \qquad (311)$$

where the proof of the last equality is left as an exercise to the reader [6].

---

6. Hint: use Eq. (104) to deduce that

$$\frac{(Ia)_{\overline{n}|i}}{a_{\overline{n}|i}} = \frac{\ddot{a}_{\overline{n}|i}}{i\, a_{\overline{n}|i}} - \frac{n\, v^n}{i\, a_{\overline{n}|i}} \qquad (312)$$

and conclude using Eqs (64) and (71).

89

Note that Eq. (311) is often rewritten in the following form:

$$D_{\text{AnnImm}}^{\text{Mac}} = \frac{1}{1-v} - \frac{n\,v^n}{1-v^n}. \tag{313}$$

Let us now consider the case of an infinite time horizon. The **Macaulay duration of a perpetuity immediate** satisfies:

$$D_{\text{PerpImm}}^{\text{Mac}} = \frac{\sum_{k=1}^{+\infty} k\,v^k}{\sum_{k=1}^{+\infty} v^k} = \frac{(Ia)_{\overline{\infty}|i}}{a_{\overline{\infty}|i}} = \frac{1}{d} = \frac{1+i}{i} = \frac{1}{1-v}, \tag{314}$$

where we take $n \to +\infty$ in Eq. (311) or in Eq. (313) and where we use $\lim_{n \to +\infty} n\,v^n = 0$ when $v < 1$.

Similarly, the **Macaulay duration of a perpetuity due** satisfies

$$D_{\text{PerpDue}}^{\text{Mac}} = \frac{\sum_{k=0}^{+\infty} k\,v^k}{\sum_{k=0}^{+\infty} v^k} = \frac{(Ia)_{\overline{\infty}|i}}{\ddot{a}_{\overline{\infty}|i}} = \frac{1}{i}. \tag{315}$$

6.6 Increasing Annuities and Perpetuities

Let us now consider the case of a financial product with geometrically increasing cash flows: $CF_{t_k} = (1+g)^k$, where $t_k = k$ for all $k \leq n$. Then, the **Macaulay duration of this geometric annuity** can be computed as follows:

$$D_{\text{gAnn}}^{\text{Mac}} = \frac{\sum_{k=1}^{n} k(1+g)^k v^k}{\sum_{k=1}^{n} (1+g)^k v^k} = \frac{\sum_{k=1}^{n} k\,v_j^k}{\sum_{k=1}^{n} v_j^k} = \frac{(Ia)_{\overline{n}|j}}{a_{\overline{n}|j}} = \frac{1+j}{j} - \frac{n\,v_j^n}{1-v_j^n}, \tag{316}$$

where

$$v_j = (1+g)\,v = \frac{1+g}{1+i} = \frac{1}{1+j} \tag{317}$$

and

$$j = \frac{i-g}{1+g}, \tag{318}$$

and where the result in Eq. (316) is obtained by mimicking the result in Eq. (311). All of this is possible because we introduce the problem-specific rate j, which should not be confused with other interest rates for which the notation j is also used.

Let us now consider the case of an infinite time horizon. We compute the **Macaulay duration of a geometric perpetuity** as follows:

$$D_{\text{gPerp}}^{\text{Mac}} = \frac{\sum_{k=1}^{+\infty} k\,(1+g)^k\,v^k}{\sum_{k=1}^{+\infty} (1+g)^k\,v^k} = \frac{\sum_{k=1}^{+\infty} k\,v_j^k}{\sum_{k=1}^{+\infty} v_j^k} = \frac{(Ia)_{\overline{\infty}|j}}{a_{\overline{\infty}|j}} = \frac{1}{d_j} = \frac{1+j}{j}, \tag{319}$$

where this result follows by letting $n \to +\infty$ in Eq. (316) and from $\lim_{n \to +\infty} n\,v_j^n = 0$ when $v_j < 1$.

The previous result can be useful for computing the duration of dividend-paying stocks, for which dividend payments are assumed to be made forever and to increase geometrically. Let us now study stocks.

6.7 Stocks

There are two different ways in which stocks provide wealth to investors: through the payment of dividends and through an increase in the price of the stock itself. Because the exam FM of the SOA concentrates on bonds and similar products such as annuities and perpetuities, the introduction to the valuation of stocks that we provide here is very brief and we only examine the effect of dividends on stock prices.

A stock price P_0 can be computed at time 0 using a so-called **Dividend Discount Model**, or DDM, as follows:

$$P_0 = \frac{E(D)}{\text{CoC} - g}, \tag{320}$$

where $E(D)$ is the expected dividend of the next period, CoC is the cost of capital (the rate of return required by stockholders), and g is the growth rate of dividends.

There are several dividend discount models. The most well-known of these models is the **Gordon Growth Model**, or GGM, which sums up the present values of all of the future dividends as follows [7]:

$$P_0 = \sum_{k=1}^{+\infty} \frac{D_0 \cdot (1+g)^k}{(1+i)^k} = \frac{D_1}{i-g} = \frac{D_0 \cdot (1+g)}{i-g}. \quad (321)$$

In this equation, we discount using the **required rate of return** of the stock i, which is another way of expressing the cost of capital.

What makes the Gordon growth model a specific version of a dividend discount model is that 1) it explicitly assumes that the rate of dividend g is constant (so that this model is more appropriate for mature companies) and 2) it replaces the expected next dividend $E(D)$ with its estimated value D_1, where D_1 is provided by analysts.

Be extremely careful with the Gordon growth model: in many exercises, you are given the current dividend D_0, not the estimated dividend D_1 in one year. In that case, you should use the rightmost part of Eq. (321) that incorporates a multiplicative factor of $1+g$.

In most situations, financial analysts assume that the number of stocks is constant when they use a dividend discount model. However, there are times when it is realistic and useful to assume that new stocks will be issued. Let w be the rate of increase in the number of company shares. Let N_0 be the initial number of shares and let D_t be the total amount of dividend paid to all stockholders by the company at time t. Then, the amount DPS_t of Dividend distributed Per Share at time t is equal to

$$\text{DPS}_t = \frac{D_t}{N_0 \, (1+w)^t}, \quad (322)$$

where this property can be used to modify formulas such as that shown in Eq. (321).

---

[7]. To prove this result, you can for instance write $P_0 = D_1 \, (Ga)_{\overline{\infty}| i, g}$ and use Eq. (165).

We can use Eqs (317), (318), and (319) to compute the **Macaulay duration of a stock**:

$$D^{\text{Mac}}_{\text{Stock}} = \frac{1+i}{i-g}, \qquad (323)$$

from which we deduce the **modified duration of a stock**:

$$D^{\text{Mod}}_{\text{Stock}} = \frac{1}{i-g}, \qquad (324)$$

using Eq. (300). This last equation could have been directly derived using Eqs (298) and (321).

We can also compute (this is a recommended exercise!) the **modified convexity of a stock**:

$$C^{\text{Mod}}_{\text{Stock}} = \frac{2}{(i-g)^2}, \qquad (325)$$

using Eqs (302) and (321). When i and g are close, this convexity can very quickly become huge.

Note that all of the computations performed in this section are for **common** stocks. Companies also issue a class of securities that are stocks but whose behavior is intermediate between that of stocks and that of bonds. These securities, called **preferred** stocks, pay a guaranteed dividend. They can be evaluated for instance by setting $g = 0$ in Eq. (321), where D_1 is known at time 0 and will be paid at the end of each year in perpetuity[8]. Under such assumptions, their value is

$$P_0^{\text{Pref. Stock}} = \frac{D_1}{i}. \qquad (326)$$

6.8 Approximations

Let P, or $P(i)$, be the price of a bond when the yield is i. Further, let P', or $P(i')$, be the price of the same bond when the

---

8. Other types of preferred stocks pay dividends indexed on a floating reference such as a Libor interest rate. Their study is beyond the scope of this book.

yield is i'. Knowing any of the two values P or P', we would like to be able to derive the second value P' or P with a simple formula, without fully recomputing the price of the bond.

Using a Taylor approximation at order 1 and Eq. (298), we can easily derive an expression of P' as a function of P called the **first-order modified approximation**:

$$P' \approx P \left(1 - \Delta i \cdot D^{\text{Mod}}\right), \tag{327}$$

where $\Delta i = i' - i$.

Using a Taylor approximation at order 1 and Eq. (291), we can derive another expression for P' after a few steps [9]. This expression, which is more complicated but also more accurate than the first-order modified approximation, is called the **first-order Macaulay approximation** and is written as follows:

$$P' \approx P \left(\frac{1+i}{1+i'}\right)^{D^{\text{Mac}}} = P \left(\frac{1+i}{1+i+\Delta i}\right)^{D^{\text{Mac}}}. \tag{333}$$

Finally note that it is often interesting to compute a **per-

---

9. A sketch of the proof is as follows. Eq. (291) can be written as

$$\frac{\partial P}{\partial \delta} = -P \, D^{\text{Mac}}. \tag{328}$$

This mathematical infinitesimal expression can be reexpressed with discrete increments as follows:

$$\frac{P' - P}{\delta' - \delta} \approx -P \cdot D^{\text{Mac}}, \tag{329}$$

so that

$$\frac{P' - P}{P} \approx -D^{\text{Mac}} \cdot (\delta' - \delta) = -D^{\text{Mac}} \cdot \left(\ln(1+i') - \ln(1+i)\right), \tag{330}$$

or

$$\frac{P' - P}{P} \approx \ln\left(\frac{1+i}{1+i'}\right)^{D^{\text{Mac}}}. \tag{331}$$

Therefore,

$$\left(\frac{1+i}{1+i'}\right)^{D^{\text{Mac}}} \approx e^{\frac{P'-P}{P}} \approx 1 + \frac{P'-P}{P} = \frac{P'}{P}, \tag{332}$$

which gives the result.

centage price change, given by:

$$\text{PPC} = 100 \left(\frac{P'-P}{P}\right) = 100 \left(\frac{P'}{P} - 1\right). \tag{334}$$

6.9 Forward Rates

Consider a loan that starts in the future. This loan consists in an amount R_{T_1} borrowed at a future time $T_1 > 0$ and in an amount R_{T_2} reimbursed at a posterior future time $T_2 > T_1$. These two amounts are related by a **forward interest rate** $f(T_1, T_2)$ such that

$$R_{T_1} = \frac{R_{T_2}}{(1 + f(T_1, T_2))^{T_2 - T_1}}, \tag{335}$$

where forward rates are generalized interest rates that allow you to compound or discount money from one point in the future to another point in the future.

We also denote:

$$f(T_1, T_2) = {}_{T_2 - T_1} f_{T_1}, \tag{336}$$

where T_1, which is the right subscript of f in the right-hand-side of the equation, is the starting point in the future, and $T_2 - T_1$, which is the left subscript of f in the right-hand-side of the equation, is the amount of time needed to reach the arrival point in the future T_2.

It is possible to relate two zero-coupon bond prices using forward rates:

$$Z(0, T_2) = Z(0, T_1) \frac{1}{(1 + f(T_1, T_2))^{T_2 - T_1}}, \tag{337}$$

where $Z(0, T_i)$ is the price at time 0 of one unit of cash paid at time T_i, for $i = 1, 2$.

You might have observed that Eq. (337) seems to be reversed compared to Eq. (335). Eq. (246) helps you understand this fact, where the currency unit that is priced by $Z(0, T_1)$ corresponds to R_{T_1} in Eq. (335).

95

Eq. (337) has a very important financial meaning. It says that directly discounting from time T_2 to time 0, which is the left-hand-side of the equality, is equivalent to first discounting from time T_2 to time T_1, and then from time T_1 to time 0, which is the right-hand-side of the equality.

Using the relation between a zero-coupon bond price and a same maturity spot rate:

$$Z(0,t) = \frac{1}{(1+s_t)^t}, \tag{338}$$

which is a subcase of Eq. (244), we reformulate Eq. (337) as follows:

$$(1+s_{T_2})^{T_2} = (1+s_{T_1})^{T_1} (1+f(T_1,T_2))^{T_2-T_1}. \tag{339}$$

We see that directly compounding from time 0 to time T_2, which is the left-hand-side of the equality, is equivalent to first compounding from time 0 to time T_1, and then from time T_1 to time T_2, which is the right-hand-side of the equality. We will remember that $\frac{1}{(1+f(T_1,T_2))^{T_2-T_1}}$ is a discount factor from time T_2 to time T_1, while $(1+f(T_1,T_2))^{T_2-T_1}$ is a compound factor from time T_1 to time T_2.

It is possible to generalize Eq. (339) as follows:

$$(1+s_{T_n})^{T_n} = (1+s_{T_1})^{T_1} \prod_{i=2}^{n}(1+f(T_{i-1},T_i))^{T_i-T_{i-1}}, \tag{340}$$

which is useful when you are given a table of forward rates. When you are given annual periods and rates, so when $T_i = i$ for all i, the previous formula becomes:

$$(1+s_n)^n = (1+s_1) \prod_{i=2}^{n}(1+f(i-1,i)). \tag{341}$$

Based on Eq. (339), we can derive the following expression for an **annual effective forward rate**:

$$_{T_2-T_1}f_{T_1} = f(T_1,T_2) = \left[\frac{(1+s_{T_2})^{T_2}}{(1+s_{T_1})^{T_1}}\right]^{\frac{1}{T_2-T_1}} - 1, \tag{342}$$

as a function of spot rates.

For simplicity, we often denote by $_1f_k = f_k$ the forward rate between two consecutive times k and $k+1$. It is also worthwhile observing that a forward rate whose starting point is the current time is simply a spot rate:

$$f(0,t) = s_t. \tag{343}$$

The following formula, derived from Eq. (338), provides another useful computation trick:

$$s_1 = \frac{1}{Z(0,1)} - 1. \tag{344}$$

In the chapter dedicated to interest rate swaps, we will make an extensive use of **periodic effective forward rates**. Such a rate is defined over the period (T_1, T_2) as follows:

$$f^*(T_1, T_2) = \frac{(1+s_{T_2})^{T_2}}{(1+s_{T_1})^{T_1}} - 1, \tag{345}$$

where the exponent $\frac{1}{T_2-T_1}$ from Eq. (342) is dropped.

Finally note that the equations of this section are sometimes written with a different notation, where spot rates are denoted by r instead of s.

Chapter 7

Immunization

Compared to most other types of businesses, insurance companies have inverted production cycles and balance sheets. What do I mean by that? Well, an insurance company is paid for producing products that still do not exist. It receives premiums and, usually, *only after that*, it insures individuals or companies. Because an insurance company owes a service in advance to its policyholders, these stakeholders are in fact liabilityholders. Therefore, the core business of an insurance company can be read in the liabilities part [1] of its balance sheet, while the core business of a standard company can be read in the assets part of its balance sheet. What does an insurance company do with all of the money it collects? It invests this money in assets that can either produce cash flows, or be sold, to pay back liabilities when they come due.

Thus, a critical aspect of the **asset-liability management** of an insurance company is making sure that the assets will always be sufficient to pay back the liabilities. Written equivalently, the **surplus** of an insurance company, defined as the difference between its assets and liabilities, should always remain positive or null. Asset-liability managers do their best to make this surplus **immune**, or insensitive, to interest rate fluctuations. A first approach to achieving this goal is **cash flow matching**.

---

1. Because insurance companies are primarily financed by their policyholders, they usually do not need to issue bonds.

7.1 Cash Flow Matching

An insurance company that implements cash flow matching builds its portfolio of assets in such a way that each liability cash flow is matched by an asset cash flow. Thus, by constructing a portfolio of assets that exactly matches the portfolio of liabilities, an insurance company guarantees that its surplus is null and remains null (put differently, this surplus is immune to interest rate movements).

To illustrate how cash flow matching is implemented in practice, assume that two liability payments of L_1 and L_2 are to be made at times 1 and 2, respectively. These liability payments can be matched using two bonds that mature at times 1 and 2.

Assume that the first of these bonds pays C_1 in coupon and P_1 in principal at time 1, while the second bond pays C_2 in coupon at times 1 and 2, and P_2 in principal at time 2. We look for the quantities q_1 and q_2 of these bonds that immunize the two liability payments.

Liability matching is achieved by first computing the quantity q_2 of the second bond that allows us to match cash flows at time 2. This is obtained by setting:

$$L_2 = q_2 \left(P_2 + C_2\right), \qquad (346)$$

or

$$q_2 = \frac{L_2}{P_2 + C_2}, \qquad (347)$$

where we set $P_2 = 0$ in some exercises if an annuity is used rather than a coupon bond.

Liability matching continues by observing that buying q_2 units of the second bond entails a payment of $q_2 \, C_2$ at time 1. Therefore, only the amount $L_1 - q_2 \, C_2$ has to be matched at time 1 by investing in the first bond. We should have:

$$L_1 - q_2 \, C_2 = q_1 \left(P_1 + C_1\right), \qquad (348)$$

or

$$q_1 = \frac{L_1 - q_2 \, C_2}{P_1 + C_1}, \qquad (349)$$

which gives the number of units of the first bond that the insurance company should purchase. In some applications, we have $C_1 = 0$ if the first bond is a zero-coupon bond.

If more than two times are considered, we should proceed recursively from the last payment time to the first payment time using the procedure outlined above.

Exact matching is obtained by matching the exact future values of the cash flows. This approach is nearly impossible to implement at the level of a real-world insurance company, for two main reasons. First, it is impossible to know the exact size and timing of future liability cash flows. Second, even if it were possible, this would entail a very complex and costly investment policy.

7.2 Immunization Principles

Because cash flow matching is nearly impossible in practice, insurance companies use another approach to protect themselves from interest rate fluctuations. This approach, called **immunization**, matches the total present values, and some sensitivity indicators, of liability and asset cash flows [2].

A big advantage of immunization is that it offers many **more investment choices** than cash flow matching. Indeed, it is a much less constrained method, because the constraints are applied at the aggregate level rather than at the cash flow level.

Immunization is **independent of the shape of the yield curve**. For instance, it does not assume a flat yield curve in the general case, although some implementations based on the Macaulay or modified duration implicitly make this additional assumption. However, immunization always assumes that only **parallel shifts** of the yield curve are possible.

When *small* parallel yield curve shifts occur, the surplus of an immunized balance sheet remains positive or (approximately) null. It is only under **full immunization** that parallel yield

---

2. From a theoretical viewpoint, cash flow matching could be considered as a subcase of immunization. For clarity, we distinguish the two approaches.

curve shifts *of any size* can never produce a negative insurance company surplus. Nothing can be anticipated in the event of non-parallel yield curve shifts, which are beyond the scope of our study.

While no rebalancing is needed in the exact matching approach (as long as the future cash flows are realized as anticipated), immunization requires the rebalancing of the assets portfolio when interest rates fluctuate. Indeed, durations and other sensitivity indicators vary when interest rates vary. Specifically, the **rebalancing frequency** of assets depends on how durations vary with interest rate movements. Therefore, the rebalancing frequency is best explained by a sensitivity of sensitivity type indicator, such as the convexity.

From now on, we distinguish full immunization from Redington immunization. Both approaches satisfy the assumptions and observations made above.

As already sketched, **full immunization** produces an aggregate portfolio that is **immune to changes of any size** in the interest rate. Full immunization is achieved by **matching the price and a first order sensitivity** to the interest rate[3] of a portfolio of liabilities and of a portfolio of assets. These two conditions are necessary but not sufficient to entail full immunization. A third condition is necessary. For instance, we often also require that each liability cash flow is **both preceded and followed** by an asset cash flow.

Redington immunization produces an aggregate portfolio that is only **immune to small or medium changes** in the interest rate. Redington immunization is achieved by **matching the price and a first order sensitivity** (to the interest rate) of a portfolio of liabilities and of a portfolio of assets. An additional condition of Redington immunization is that the **convexity of assets be superior to the convexity of liabilities**. This third condition broadens the range of interest rate fluctuations for which the surplus remains positive. Be careful though that this condition is not sufficient to guarantee full immunization in

---

3. For instance, the modified duration or the Macaulay duration.

the general case.

What is the exact meaning of "an aggregate portfolio (or surplus) is immune to changes in interest rates"? As already sketched, we mean that the value of the total portfolio (or surplus) either remains the same, or perhaps increases, when the yield curve is subject to a parallel shift (when all interest rates move by the same amount). Put differently, the value of the immunized portfolio (or surplus) is at its minimum. In the case of full immunization, no parallel yield curve shift can reduce this value, which is therefore at its global minimum. In the case of Redington immunization, this minimum is only local in the sense that large parallel yield curve shifts can reduce the value of the aggregate portfolio (or surplus). Therefore, full immunization entails Redington immunization, but the converse is not necessarily true.

Remember that all of the previous discussion only holds under strong assumptions. A portfolio that is fully immunized against parallel yield curve shifts can be subject to losses should a nonparallel yield curve shift occur, so should long term interest rates move differently than short term interest rates.

7.3 A First Illustration

Let L be the value of a liability and D_L^{Mac} be its Macaulay duration. We assume that two assets are available to immunize this liability. The first asset has a value A_1 and a Macaulay duration $D_{A_1}^{\text{Mac}}$. The second asset has a value A_2 and a Macaulay duration $D_{A_2}^{\text{Mac}}$. Let N_1 and N_2 be the *numbers* (not necessarily integers) of such assets in the asset portfolio. Immunization amounts to equating the present values:

$$L = N_1 \, A_1 + N_2 \, A_2 \qquad (350)$$

and the Macaulay durations of the assets and liability portfolios:

$$L \, D_L^{\text{Mac}} = N_1 \, A_1 \, D_{A_1}^{\text{Mac}} + N_2 \, A_2 \, D_{A_2}^{\text{Mac}}, \qquad (351)$$

where we use the fact that duration is a linear indicator.

Denoting by $x_1 = N_1 \, A_1$ and $x_2 = N_2 \, A_2$ the *amounts* invested in the two assets, the above two immunization conditions can be rewritten as follows:

$$L = x_1 + x_2 \tag{352}$$

and

$$L \, D_L^{\text{Mac}} = x_1 \, D_{A_1}^{\text{Mac}} + x_2 \, D_{A_2}^{\text{Mac}}. \tag{353}$$

A third alternative representation of the immunization constraints can be derived using *proportions*. Let $q_1 = N_1 \, \frac{A_1}{L}$ and $q_2 = N_2 \, \frac{A_2}{L}$ be the proportions invested in the two assets. Then,

$$1 = q_1 + q_2 \tag{354}$$

and

$$D_L^{\text{Mac}} = q_1 \, D_{A_1}^{\text{Mac}} + q_2 \, D_{A_2}^{\text{Mac}}. \tag{355}$$

The next illustrations replace the Macaulay duration with the first derivative (computed with respect to the yield) of asset and liability prices for the construction of first-order constraints. The two approaches are completely equivalent.

7.4 A Second Illustration

Consider a liability that consists in a unique cash flow C made at time n. We can immunize this liability using two assets. The first asset pays a unique cash flow A_1 at time $n - h$, while the second asset pays a unique cash flow A_2 at time $n + h$. We look for the values of A_1 and A_2 that produce an immunized surplus.

All cash flows are discounted using a unique yield y, which amounts to assuming a flat yield curve. First, we cancel the **surplus** $\mathcal{S}$, defined as the difference between the present value of the assets and the present value of the liabilities, and we write:

$$\mathcal{S}(y) = \frac{A_1}{(1+y)^{n-h}} + \frac{A_2}{(1+y)^{n+h}} - \frac{C}{(1+y)^n} = 0. \tag{356}$$

Next, we construct a first-order constraint by also canceling the first derivative of the surplus:

$$S'(y) = -\frac{(n-h)\,A_1}{(1+y)^{n-h+1}} - \frac{(n+h)\,A_2}{(1+y)^{n+h+1}} + \frac{n\,C}{(1+y)^{n+1}} = 0, \quad (357)$$

which is equivalent to canceling the Macaulay duration or the modified duration of the surplus.

Next, we multiply Eqs (356) and (357) by $(1+y)^{n+h}$ and $(1+y)^{n+h+1}$, respectively. We obtain the equivalent system of equations:

$$A_1\,(1+y)^{2h} + A_2 = C\,(1+y)^h \qquad (358)$$

and

$$(n-h)\,A_1\,(1+y)^{2h} + (n+h)\,A_2 = n\,C\,(1+y)^h. \qquad (359)$$

Subtracting n times Eq. (358) from Eq. (359), and dividing by $-h$, we also obtain:

$$A_1\,(1+y)^{2h} - A_2 = 0. \qquad (360)$$

The solutions to Eqs (356) and (357) are the same as the solutions to Eqs (358) and (360). You can check that they are equal to

$$A_1 = \frac{C}{2\,(1+y)^h} \qquad (361)$$

and

$$A_2 = \frac{C\,(1+y)^h}{2}, \qquad (362)$$

where the asset cash flows A_1 and A_2 are simply equal to half the liability cash flow C, discounted or compounded for the period of time separating cash flows.

7.5 A Third Illustration

When the time h_1 between the first asset cash flow and the liability cash flow is different from the time h_2 between the liability cash flow and the second asset cash flow, the equation that

describes the cancellation of the surplus becomes:
$$S(y) = \frac{A_1}{(1+y)^{n-h_1}} + \frac{A_2}{(1+y)^{n+h_2}} - \frac{C}{(1+y)^n} = 0, \quad (363)$$

while the first-order constraint on the surplus is written as follows:
$$S'(y) = -\frac{(n-h_1)\,A_1}{(1+y)^{n-h_1+1}} - \frac{(n+h_2)\,A_2}{(1+y)^{n+h_2+1}} + \frac{n\,C}{(1+y)^{n+1}} = 0. \quad (364)$$

These two equations are equivalent to the following pair of equations:
$$A_1\,(1+y)^{h_1} + A_2\,(1+y)^{-h_2} - C = 0, \quad (365)$$

and
$$h_1\,A_1\,(1+y)^{h_1} - h_2\,A_2\,(1+y)^{-h_2} = 0, \quad (366)$$

which generalize Eqs (358) and (360).

You can check that the solutions to this pair of equations are
$$A_1 = \frac{h_2}{h_1 + h_2}\,\frac{C}{(1+y)^{h_1}} \quad (367)$$

and
$$A_2 = \frac{h_1}{h_1 + h_2}\,C\,(1+y)^{h_2}. \quad (368)$$

7.6 A Further Generalization

Let us now look at the general case where K asset cash flows and H liability cash flows are considered. The size of an asset cash flow that occurs at time a_k is denoted by A_k, where $k = 1, \cdots, K$. Similarly, the size of a liability cash flow that occurs at time l_h is denoted by L_h, where $h = 1, \cdots, H$.

The cancellation of the surplus is written as follows:
$$S(y) = \sum_{k=1}^{K} \frac{A_k}{(1+y)^{a_k}} - \sum_{h=1}^{H} \frac{L_h}{(1+y)^{l_h}} = 0, \quad (369)$$

while the cancellation of the first derivative of the surplus is written as

$$S'(y) = -\sum_{k=1}^{K} \frac{a_k \, A_k}{(1+y)^{a_k+1}} + \sum_{h=1}^{H} \frac{l_h \, L_h}{(1+y)^{l_h+1}} = 0, \qquad (370)$$

where, again, this is equivalent to canceling the Macaulay duration or the modified duration of the surplus.

For the company to be immunized in the Redington sense, an additional constraint is required. The second derivative of the surplus should be greater than zero, which we write as follows:

$$S''(y) = \sum_{k=1}^{K} \frac{a_k \, (a_k+1) \, A_k}{(1+y)^{a_k+2}} - \sum_{h=1}^{H} \frac{l_h \, (l_h+1) \, L_h}{(1+y)^{l_h+2}} \geq 0. \qquad (371)$$

Chapter 8

Interest Rate Swaps

Interest rate swaps are agreements where two **counterparties** agree to exchange series of cash flows indexed on fixed and floating interest rates. As such, these contracts are traded over the counter, meaning that they are intermediated by financial institutions such as banks, but are not sold on organized exchanges. These contracts are particular examples of derivatives, namely forward contracts, where the counterparties bind themselves to perform a given action (here, paying fixed or floating cash flows).

Cash flow payments are made at times called **settlement dates**, while the **settlement period** is the time lag between two payments. In most applications, we assume that the settlement period is annual or semiannual.

The **term**, or **tenor**, of the swap, is the time length of the swap. We denote by n the number of cash flows exchanged during the life of the swap. As before, when payments are annual (so when the settlement period is equal to one year), the term is usually equal to the number of payments, except if exotic covenants are added.

A swap consists in exchanging a series of fixed cash flows against a series of floating cash flows: the series of fixed cash flows is called the **fixed leg** of the swap, while the series of floating cash flows is called the **floating leg** of the swap.

How do we distinguish the counterparty that makes fixed payments and receives floating payments from the counterparty that makes floating payments and receives fixed payments? We say that an investor enters into a **payer swap** when she makes fixed payments and receives floating payments. Similarly, we say that an investor enters into a **receiver swap** when he receives fixed payments and makes floating payments. Therefore, the naming convention is based on the fixed leg of the swap.

A so-called **notional amount** Q is used to compute the value of the fixed and floating payments. However, be careful that this amount is not paid at the maturity. In this sense, it is different from a principal amount that is paid at the maturity of a bond[1].

The fixed leg corresponds to a series of payments proportional to the notional amount Q and computed based on a fixed rate called the **swap rate**, which we denote by R. Each cash flow CF^{fi} of the fixed leg is equal to[2]:

$$CF^{fi} = R \cdot Q \qquad (373)$$

where you can take a look at Fig. 19 for a cash flow timeline.

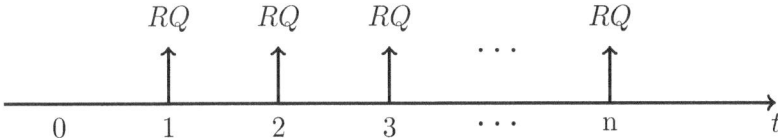

Figure 19 – Cash Flow Timeline of the Fixed Leg of a Swap

The floating leg corresponds to a series of payments proportional to the notional amount Q and computed based on a floating

---

1. In currency swaps, final principal payments are made in two different currencies. This is for your information only, because the study of currency swaps is out of the scope of this book.

2. Note that swap rates are sometimes expressed as excesses with respect to a Government reference. For instance, let y be an effective Government bond (or treasury) yield. Then, a swap rate R can be expressed as the sum of such a yield and of a **swap spread** SS as follows:

$$R = y + SS. \qquad (372)$$

rate published by external sources and regularly updated. The kth floating cash flow CF_k^{fl} is equal to:

$$\text{CF}_k^{\text{fl}} = L_{k-1} \cdot Q, \qquad (374)$$

where L_{k-1} is a floating rate for the period $(k-1, k)$. L_{k-1} is usually prefixed, meaning that it is published at the beginning of the period, so at time $k-1$. Therefore, apart from the first cash flow that is usually known at time 0, all of the other cash flows are unknown at that time. See Fig. 20 for a cash flow timeline of the floating leg of a swap, where the first arrow is straight to indicate that the first cash flow is known at time 0, and the other arrows are wavy to indicate that the other cash flows are unknown at time 0.

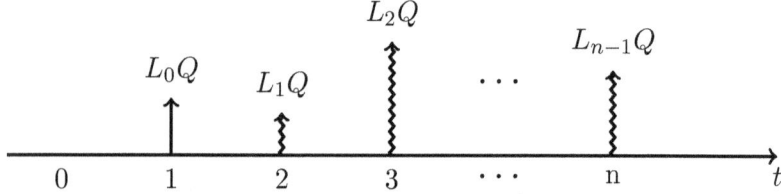

Figure 20 – Cash Flow Timeline of the Floating Leg of a Swap

Sometimes a constant fixed rate c is added to the floating payments, producing floating payments of the following kind:

$$\text{CF}_k^{\text{fl}} = (L_{k-1} + c) \cdot Q. \qquad (375)$$

Typical floating rates are the Libor (for London InterBank Offered Rate) and the Euribor (for Euro InterBank Offered Rate). These rates are the rates that prevail for loans between top international banks. Libors and Euribors exist in several currencies, including the US dollar. For instance, a Libor interest rate exists for loans denominated in yens. The prime rate and the federal funds rate (discussed in the next chapter) are other possible floating references in the US.

It is important to understand that CF^{fi} and CF_k^{fl} are not actually paid. In fact, they are **netted**, and only their difference is paid by one counterparty to the other. In practice, the following **net swap payment** NSP_k:

$$\text{NSP}_k = \pm \left(\text{CF}_k^{\text{fi}} - \text{CF}_k^{\text{fl}} \right) \qquad (376)$$

is *made* at time k. In the case of the counterparty who has entered a payer swap (the counterparty that pays the fixed payment and receives the floating payment, as mentioned beforehand), the '$\pm$' sign is a '+', while it is a '-' for the counterparty who has entered a receiver swap.

Most often, investors enter into swaps to modify the pattern of a series of cash flows that they either receive or pay. For instance, a company that borrowed money via a floating rate bond has to pay floating interest rate payments. It may want to pay fixed interest rate payments instead. To do so, it will enter into a payer swap: it will pay fixed payments to an unknown counterparty through its bank and it will pay zero floating payments in total because the floating payments received from the swap counterparty offset the floating payments that must be made on the floating rate bond.

To describe the aggregate situation of an investor, we compute the **net interest payment** NIP_k made at time k as the sum of NCFL_k, which is the net cash flow paid on the loan (or multiple loans) at time k, and of NSP_k, which is the net swap payment made at time k:

$$\text{NIP}_k = \text{NCFL}_k + \text{NSP}_k. \qquad (377)$$

Denote by $\mathcal{V}_0^r$ the **initial market value** of a receiver swap. This is the difference between the present value of the fixed leg that is received by the investor and the present value of the floating leg that is paid by the investor. By an appropriate choice of swap rate R, the initial market value of a swap is often set equal to zero [3], so that we have [4]:

$$\mathcal{V}_0^r = Q\, R \sum_{k=1}^{n} P_{t_k} - Q \sum_{k=1}^{n} f^*(t_{k-1}, t_k)\, P_{t_k} = 0, \qquad (379)$$

---

3. In the rare situation where this is not the case, an upfront amount is exchanged between the counterparties to offset this initial market value.

4. The initial market value $\mathcal{V}_0^p$ of a payer swap is also most often null and

where P_{t_k} is the discount factor for a cash flow that is made at time t_k. In this equation, $R\, P_{t_k}$ is the present value of a fixed payment made at time k, while $f^*(t_{k-1}, t_k)\, P_{t_k}$ is the present value of a floating payment made at time k, where $f^*(t_{k-1}, t_k)$ is the periodic effective forward rate for the period $(k-1, k)$ defined in Eq. (345).

From this condition, we readily derive the swap rate R, which is equal to

$$R = \frac{\sum_{k=1}^{n} f^*(t_{k-1}, t_k)\, P_{t_k}}{\sum_{k=1}^{n} P_{t_k}} = \frac{\sum_{k=1}^{n} \left(\frac{(1+s_{t_k})^{t_k}}{(1+s_{t_{k-1}})^{t_{k-1}}} - 1 \right) \frac{1}{(1+s_{t_k})^{t_k}}}{\sum_{k=1}^{n} \frac{1}{(1+s_{t_k})^{t_k}}},$$
(380)

where s_{t_k} is the spot rate for the period $(0, t_k)$. This spot rate is also sometimes denoted as r_{t_k}.

For *annual* settlement periods, the periodic effective forward rates become annual effective forward rates, and we have for the first year: $f^*(t_0, t_1) = f^*(0, 1) = f(0, 1) = s_1$.

We can write:

$$1 = \sum_{k=1}^{n} f^*(t_{k-1}, t_k)\, P_{t_k} + P_{t_n}, \qquad (381)$$

which says that the value of one dollar today is equal to the sum of the values of future interest payments whose current best estimate is f^*_{k-1} and of the value of a principal payment of one dollar made at the future time t_n. These values are computed by discounting with the factors P_{t_k} and P_{t_n}, respectively. This formula is nothing but a bond pricing formula.

We can use Eq (381) to rewrite the swap rate of Eq. (380) as

---

can be written as

$$V_0^p = Q \sum_{k=1}^{n} f^*(t_{k-1}, t_k)\, P_{t_k} - Q\, R \sum_{k=1}^{n} P_{t_k} = 0. \qquad (378)$$

follows:
$$R = \frac{1 - P_{t_n}}{\sum_{k=1}^{n} P_{t_k}} = \frac{1 - \frac{1}{(1+s_{t_n})^{t_n}}}{\sum_{k=1}^{n} \frac{1}{(1+s_{t_k})^{t_k}}}, \qquad (382)$$

which is much quicker to compute.

After time 0, the **market value** of the swap becomes non-null. This is because the swap rate, which is contractual, does not change, whereas the state of the market evolves and promises different future expected floating cash flows. For instance, if the yield curve is bumped upwards, the value of the floating leg becomes higher than it previously was and in particular larger than the value of the fixed leg, yielding a negative market value for a receiver swap and a positive market value for a payer swap. Similarly, if the yield curve is bumped downwards, the value of the floating leg becomes lower than it previously was and in particular smaller than the value of the fixed leg, yielding a positive market value for a receiver swap and a negative market value for a payer swap.

Specifically, denote by t a reevaluation time assumed larger than 0. Then, the market value at time t of a receiver swap $\mathcal{V}_t^r$ is computed as follows:

$$\mathcal{V}_t^r = Q\,R\sum_{k=k_t}^{n} P'_{t_k} - Q\sum_{k=k_t}^{n} f'^{*}(t_{k-1}, t_k)\,P'_{t_k}, \qquad (383)$$

or as

$$\mathcal{V}_t^r = \sum_{k=k_t}^{n} Q\left(R - f'^{*}(t_{k-1}, t_k)\right) P'_{t_k}, \qquad (384)$$

where k_t is the time of the first cash flow that occurs immediately after time t. This notation allows us to perform valuations between payment times, although in the exercises you will most often encounter the situation where t is a cash flow payment time. The prime in the discount factor P'_{t_k} means that it is computed using the new yield curve at time t, where discounting is performed from time t_k to time t. Similarly, the periodic effective forward rate $f'^{*}(t_{k-1}, t_k)$ is computed using the new yield curve at time t and gives the expected interest rate for the period (t_{k-1}, t_k) viewed from time t.

The market value at time t of a payer swap $\mathcal{V}_t^p$ is computed in the same way:

$$\mathcal{V}_t^p = Q \sum_{k=k_t}^{n} f'^*(t_{k-1}, t_k) P'_{t_k} - Q R \sum_{k=k_t}^{n} P'_{t_k}, \qquad (385)$$

so that

$$\mathcal{V}_t^p = \sum_{k=k_t}^{n} Q \left(f'^*(t_{k-1}, t_k) - R \right) P'_{t_k}. \qquad (386)$$

Note that the following condition is always satisfied:

$$\mathcal{V}_t^r = -\mathcal{V}_t^p, \qquad (387)$$

so that when a positive market value is generated for any of the counterparties of a swap due to an interest rate market move, a negative market value of the same size is generated for the other counterparty of the swap.

A **deferred swap** makes its first payment at a date later than time t_1. Written equivalently, the time lag between the issuance date and the first settlement date is larger than the settlement period.

Assume for instance that the first payment of a swap occurs at time t_m, but that the swap maturity is still t_n. Then, we can compute the swap rate of this deferred swap as follows:

$$R = \frac{\sum_{k=m}^{n} f^*(t_{k-1}, t_k) P_{t_k}}{\sum_{k=m}^{n} P_{t_k}} = \frac{\sum_{k=m}^{n} \left(\frac{(1+s_{t_k})^{t_k}}{(1+s_{t_{k-1}})^{t_{k-1}}} - 1 \right) \frac{1}{(1+s_{t_k})^{t_k}}}{\sum_{k=m}^{n} \frac{1}{(1+s_{t_k})^{t_k}}}. \qquad (388)$$

To simplify the above formula, we need to generalize Eq. (381). We obtain:

$$P_{t_{m-1}} = \sum_{k=m}^{n} f^*(t_{k-1}, t_k) P_{t_k} + P_{t_n}, \qquad (389)$$

where we express in two different ways the value at time 0 of one dollar received at time t_{m-1}. The left-hand-side of the equation provides a direct pricing of this future dollar using the discount factor $P_{t_{m-1}}$. The right-hand-side of the equation simply says that a dollar paid at time $m-1$ is equivalent to interest payments made from time t_m to time t_n plus a principal repayment at time t_n. All of these payments should be first discounted from time t_k (with k ranging between m and n) or t_n to time t_{m-1}, and then from time t_{m-1} to time 0. The right-hand-side of the equation directly directly discounts these payments from time t_k or t_n to time 0 using the discount factors P_{t_k} and P_{t_n}.

Plugging Eq. (389) into Eq. (388), we obtain:

$$R = \frac{P_{t_{m-1}} - P_{t_n}}{\sum_{k=m}^{n} P_{t_k}} = \frac{\frac{1}{(1+s_{t_{m-1}})^{t_{m-1}}} - \frac{1}{(1+s_{t_n})^{t_n}}}{\sum_{k=m}^{n} \frac{1}{(1+s_{t_k})^{t_k}}}, \quad (390)$$

which is much simpler than Eq. (388).

Consider for instance a swap that is deferred for two years and that has annual settlement periods. No exchanges of cash flows occur during the first two years. Thus, the first cash flow occurs at time $m = 3$. Therefore, the quantity $P_{t_{m-1}}$ to use in the numerator of Eq. (390) is P_2. For a swap that is not deferred, the first cash flow occurs at time $m = 1$. Thus, $P_{t_{m-1}}$ becomes $P_0 = 1$ and we recover Eq. (382).

The previous swap rate formulas were independent of the notional amount Q, which was assumed constant. This is not anymore the case when we allow the notional amount to vary.

A first example of a swap with a varying notional amount is an **amortizing swap**. For this swap, the notional amount decreases with time in the same way as the outstanding amount of a loan. Another example of a swap with a varying notional amount is an **accreting swap**, where now the notional amount increases.

When we replace Q with a varying Q_t, the swap rate formula becomes:

$$R = \frac{\sum_{k=1}^{n} f^*(t_{k-1}, t_k)\, Q_{t_k}\, P_{t_k}}{\sum_{k=1}^{n} Q_{t_k}\, P_{t_k}}, \qquad (391)$$

and the notional amount is not simplified out.

For a swap that has both a varying nominal amount and deferred payments, we can compute the swap rate by combining Eqs (388) and (391) as follows:

$$R = \frac{\sum_{k=m}^{n} f^*(t_{k-1}, t_k)\, Q_{t_k}\, P_{t_k}}{\sum_{k=m}^{n} Q_{t_k}\, P_{t_k}}. \qquad (392)$$

Chapter 9

Determinants of Interest Rates

We start by studying the Federal Reserve System, which is the most prominent central bank worldwide. Then, we briefly examine the pricing of Treasuries. Finally, we discuss the nature and functioning of interest rates in a broad and general way.

9.1 Federal Reserve System

The federal reserve system is comprised of a **Board of Governors**, a **Federal Open Market Committee** (FOMC), and twelve **federal reserve banks**. The Board of Governors is composed of seven members chosen by the President of the U.S. and supervises the whole system. The FOMC has twelve voting members: the seven Board of Governors members, the president of the federal reserve bank of New York, and four of the presidents of the other federal reserve banks (who serve on a rotating basis). The FOMC defines the U.S. monetary policy and sets target interest rates [1]. The twelve federal reserve banks not only implement the monetary policy, but also have a broad variety of roles, such as ensuring a proper functioning of the payment system.

---

1. In many countries, the committee that performs such duties is called the Monetary Policy Committee.

Banks should maintain **required reserve balances** at the Federal Reserve. These reserve requirements are computed using **required reserve ratios**. Banks that lack Federal Reserve balances borrow money from banks with an excess of cash in their reserve accounts. The interest rate charged for such *interbank* operations is the **federal funds rate**.

Note that the federal funds rate is *not* the interest rate that is charged by the Federal Reserve when it lends money to banks. The federal funds rate used by banks in their joint lending and borrowing operations is strongly influenced but *not* set by the Federal Reserve.

While the federal funds rate is the actual rate used for interbank reserves lending, the **Federal Funds Target Rate** is defined as the target rate that the FOMC would like to see applied in the implementation of such operations. The Federal Reserve can use several tools to induce a convergence of the federal funds rate towards the federal funds target rate. It can modify required reserve ratios (this is rarely done), it can conduct open market operations (buying or selling treasuries outright or using repurchase agreements), and it can modify the discount rate that is discussed hereafter.

Banks that have a shortage of liquidity can directly borrow funds from the Federal Reserve via the so-called **discount window**. The rate at which they borrow from the Fed is the **discount rate**[2]. This rate should not be mistaken with the rate d that we extensively discussed in this text and that can be used for the pricing of any loan or bond.

The discount rate should not be confused with the **prime rate**. The prime rate is the lowest rate that banks offer to their best customers. It is not chosen by the Federal Reserve but it is computed by adding a margin (often maintained constant by banks) to the federal funds rate.

In conducting its monetary policy, the Federal Reserve aims at achieving three statutory goals: facilitating a **maximum level**

---

2. In fact, the Fed uses a primary rate, a secondary rate, and a seasonal rate when conducting discount window lending. Specifically, the discount rate is the primary rate.

of employment, contributing to the **stability of prices**, and maintaining **reasonable long-term interest rates**.

A classic explanation of the link between the federal funds rate and other interest rates is as follows. If the federal funds rate increases, it becomes more interesting for banks to maintain surpluses in their reserve accounts. Therefore, they have less money to lend to customers and they write fewer loans. In turns, this shortage in available money entails an increase in the interest rates of loans.

A similar argument can be developed to explain how decreases in the federal funds rate entail decreases in the interest rates associated with loans. In passing, we have observed that an increase in the federal funds rates entails a rarefaction of the cash available in the economy, while a decrease in the federal funds rate entails an expansion of the cash available in the economy.

The Fed puts a downward pressure on the federal funds rate by buying Treasuries. This is done by increasing the size of the balance sheet of the Fed, so by creating money. This increase in available liquidity hopefully leads to more investments and to a reduction of the unemployment rate, but also sometimes to an overheating of the system and to an increase in the inflation rate.

When the Federal Reserve buys Treasuries, it receives coupons from the U.S. Government. Part of these coupons go back to the federal state after regulatory dividends are extracted. Note that a decrease in interest rates makes a country more competitive because this country's currency is depreciated as a consequence of the increase in available money. Finally, a decrease in interest rates also eases further borrowing by a government by reducing the total amount of coupons that should be paid each year. While many of the consequences of reducing interest rates can be positive in the absence of inflation, increasing too much the total amount of public debt can be dangerous in the long run.

9.2 Treasuries

The U.S. government issues debt in the form of **Treasury-Bills (or T-bills)**, **Treasury Notes (or T-notes)**, and **Trea-**

sury Bonds (or T-bonds). T-bills are zero-coupon bonds that are issued with maturities of 4, 8, 13, 26, and 52 weeks. T-notes and T-bonds pay semiannual coupons. T-notes have maturities of 2, 3, 5, 7, and 10 years, while T-bonds have the longest maturities of 20 and 30 years. The Canadian government also issues zero coupon bonds and classic bonds that pay semiannual coupons and have a maximum maturity of 30 years [3].

Because U.S. and Canadian T-bills are zero-coupon bonds, the yield that they provide takes the form of an increase in the sum that is paid back to investors at the maturity. In practice, let M_0 be the amount lent at time 0 by an investor to the U.S. or Canadian government via a T-bill. The amount M_T that is paid back at time T to the investor is larger than M_0 and the yield j of such a financial product solves the following equation:

$$M_T = M_0 \cdot (1+j)^{\frac{T}{365}}, \qquad (393)$$

where T is a number of days.

Be aware of the following market conventions. The **quoted rate of U.S. T-bills** is defined as follows:

$$QR^{US} = \frac{360}{T} \cdot \frac{M_T - M_0}{M_T}, \qquad (394)$$

while the **quoted rate of Canadian T-bills** takes the following form:

$$QR^{CND} = \frac{365}{T} \cdot \frac{M_T - M_0}{M_0}. \qquad (395)$$

The quantity $M_T - M_0$ is called the **dollar amount of interest**. In classic conditions, the rates defined above satisfy:

$$j > QR^{CND} > QR^{US}. \qquad (396)$$

9.3 Interest Rates

Government bonds returns are sometimes called **risk-free** returns. However, these returns are not at all risky, at least from

---

[3]. For illustration, the French government issues bonds that pay annual coupons and that have a maximum maturity of 50 years.

the viewpoint of the purchasing power of an investor. Indeed, the only guarantee that these bonds offer is to bring a prefixed flow of currency units. In this sense, their returns are **nominal returns**. If the economy suffers from a high level of **inflation**, these cash flows will bring to their recipient a lowered consumption capacity. Therefore, nominal Government bond returns are not risk-free, at least in terms of inflation/deflation risk.

A bond whose coupons are indexed to inflation is called a **real return bond**. It provides real returns as opposed to nominal returns. Real return bonds are offered by few governments, such as the U.S. and the French governments. These governments are able to pay inflation-linked coupons because they collect taxes that themselves increase with inflation. In the U.S., these products are called **Treasury Inflation-Protected Securities**, or **TIPS**. The face value of TIPS is proportional to a **Consumer Price Index**, or **CPI**. The constant coupon rate applied to this fluctuating face value produces coupon amounts that increase with inflation. Inflation-indexed securities can be very useful investment assets for pension funds, in order to maintain the purchasing power of their pensioners.

It may be risky for a company to issue inflation-linked bonds whose coupons are linked with the overall level of inflation. Indeed, if there is no inflation in the sector where a company operates, its income will not increase (assuming a stable level of sales), while the company will have to pay coupons that increase due to the inflation that prevails in the whole economy.

Coming back to the beginning of this section, what did we mean when we wrote that Government bonds returns are risk-free? We meant that these returns are **credit-risk-free**. It is classically assumed [4] that the U.S., Canadian, UK, German, ..., governments will never go bankrupt. In this sense, and in this sense only, the governments of the wealthiest nations issue nominal bonds that are called risk-free. Clearly, an investor exposes herself to a higher level of credit risk by investing in the bonds issued by less developed countries, or in the bonds issued by

---

4. At least at the time at which these lines are written.

provinces or municipalities. Credit risk is also taken by investing in the bonds issued by companies. Credit risk should be compensated by a **credit risk premium**.

When a company bankrupts, its assets are sold and the proceeds of the sale are distributed back to liabilityholders. Unfortunately, these proceeds are usually not sufficient to fully reimburse liabilityholders, due to the existence of bankruptcy costs. The **recovery rate** R is defined as the proportion of principal that is recovered after a bankruptcy. Let p be the bankruptcy probability and $Z(0,T)$ be the discount factor that we apply to a payment of principal P due at time T. Then, the present value of this payment of principal, taking into account credit risk, is

$$(1-p)\,P\,Z(0,T) + p\,R\,P\,Z(0,T) = (1-p+p\,R)\,P\,Z(0,T), \quad (397)$$

where the left-hand-side of this equation reads as the sum of the discounted value of the principal amount when default does not occur and of the discounted value of what is recovered from the principal amount when default occurs. Eq. (397) is a first, naive, corporate bond pricing formula.

Credit risk is not the only risk that affects corporate bond yields. Liquidity risk is another important risk and should be compensated by a **liquidity risk premium**[5]. Consider for instance two companies that issue bonds in the same currency and that have the same level of credit risk. Further assume that the bonds have identical coupons, maturity, and so on, but that the only difference is that the bonds of the first company are less liquid than the bonds of the second company. Then, the less liquid bonds should offer a higher yield to investors than the more liquid bonds, in order to compensate them for taking this additional risk.

Finally, also observe that, all else kept constant, bonds with higher maturities offer higher yields. This yield premium compensates investors for the additional uncertainty associated with holding a bond for a long time. Therefore, a **maturity risk premium** should be taken into account when assessing bond yields.

---

5. This premium is often negligible in the case of government bonds.

From the previous discussion, we deduce that any interest rate j can be decomposed into the following constituents:

$$j = r + s + i_e + i_u + l + m + \cdots, \qquad (398)$$

where r is a "pure" interest rate component, which can be interpreted as a compensation for deferred consumption or as a pure utility measure of moving money backward or forward in time, s is a credit risk premium, i_e is a premium for expected inflation, i_u is a premium for unexpected inflation, l is a liquidity risk premium, m is a maturity risk premium, and so on.

Beware that in our formulation of Eq. (398), s is a pure credit risk premium. However, l is sometimes incorporated into s in other books or papers. Further note that Eq. (398) is appropriate in the general case of a **loan without inflation protection**. Finally, Eq. (398) is often expressed in a multiplicative and more accurate form:

$$1 + j = (1 + r) \prod_{k=1}^{n} (1 + \pi_k), \qquad (399)$$

where π_k is the risk premium associated with the kth risk and n risks are taken into account in this example.

In the case of a **loan that is protected against inflation**, the following rate is paid to investors:

$$j = u + i_a, \qquad (400)$$

where i_a is the actual inflation rate (this is *not* a risk premium) that is added to a contractual rate u. This latter interest rate, which is a real interest rate, can be decomposed as follows:

$$u = r - c, \qquad (401)$$

where r is the "pure" interest rate discussed above and c is a premium charged by the lender in exchange for providing a protection against inflation.

We conclude this chapter by observing that several **yield curve shapes** exist. The most classic shapes are **increasing**,

decreasing (or **inverted**), **humped**, and **flat**. A flat yield curve shape corresponds to the situation considered in most of this book, where a unique interest rate is used for discounting cash flows of all maturities.

There are several theories that explain yield curve shapes. According to the **liquidity preference or opportunity cost theory**, lending is preferred in the short term because the money that is redeemed can be quickly reused to benefit from other investment opportunities and a positive maturity premium [6] should be added to long term yields. Therefore, this first theory is most often useful to explain increasing interest rate term structures.

According to the **market segmentation theory**, investors are characterized by different needs. While some investors need to invest in the short term, other investors need to invest in the long term. The proportions of both types of investors can vary with time, which explains why some segments of the yield curve are sometimes preferred (driving the corresponding rates down) and sometimes not. The **preferred habitat theory** is very similar to the market segmentation theory, but does not assume that investors strictly stick to particular investment terms. Finally, the **expectations theory** considers that long term rates contain information about spot short term rates and forward short term rates (see for instance Eq. (341)) and that forward short term rates are indicators of the actual short term rates observed in the future. According to this theory, the shape of the yield curve is an indicator of the expectation by market participants of the future state of the economy. For instance, an increasing yield curve corresponds to an expectation of higher future interest rates. The last three theories are consistent with any shape of the yield curve.

---

6. This risk premium differs from the maturity premium discussed beforehand that was a compensation for the additional uncertainty associated with long maturity bonds.

Appendix

The tables provided in this appendix give hints for solving the series of exercises made freely available by the Society of Actuaries. They can be read as follows: Ex. 1, or Exercise 1, is solved using Eq. 29 and 45, or Eqs 29 and 45, from this book. T1 is the first table and C7 refers to general discussions in the seventh chapter. / means the exercise has been suppressed by the SOA.

Ex.	Eq.	Ex.	Eq.	Ex.	Eq.
1	29, 45	26	57, 211, 243	51	347, 349
2	186	27	13	52	64, 347, 349
3	10, 29	28	27, 223, 224	53	247, 335
4	/	29	184	54	242, 277, 279
5	281	30	238	55	275, 277, 279
6	68, 104	31	171	56	242, 278, 279
7	66, 109	32	60	57	239, 278, 279, 280
8	/	33	244	58	/
9	64	34	234, 244	59	313, 354, 355
10	236	35	289	60	156
11	68, 162	36	314	61	55
12	35, 39	37	319	62	259, 269
13	55	38	/	63	64, 212, 224
14	64, 172	39	/	64	64, 216
15	6, 64	40	/	65	307, 308
16	215	41	/	66	300, 327
17	66	42	/	67	342
18	16, 30, 104	43	/	68	306
19	281, 283, 284	44	/	69	347, 349
20	25	45	281, 283, 284	70	C7
21	59	46	T4, T5	71	361, 362
22	236	47	16, 65	72	365, 366
23	24, 25	48	75	73	367, 368
24	/	49	60, 61	74	236, 239, 241, 242
25	25, 64, 68	50	/	75	213, 228

Table 7 – Exercises and their Related Equations, Tables, and Chapters.

Ex.	Eq.	Ex.	Eq.	Ex.	Eq.
76	236, 239, 241	101	70, 118	126	C7
77	35	102	156, 163	127	369, 370, 371
78	13, 283	103	16, 170	128	369, 370
79	13, 50	104	91, 175	129	369, 370, 371
80	/	105	13, 55	130	369, 370
81	64, 108, 207, 222	106	21, 206, 229	131	289, 295
82	288	107	223	132	246
83	283	108	/	133	15
84	70, 75	109	30, 228	134	64, 68
85	187	110	21, 206, 229	135	10, 51
86	64, 104	111	246	136	28, 68
87	13, 66	112	206, 223, 224	137	16, 66, 75
88	213, 228	113	16, 236, 242	138	91, 92
89	13, 66	114	30, 236	139	1, 63, 64
90	236, 239, 241	115	236, 242	140	4, 66
91	278, 279	116	3, 30	141	34, 70, 104
92	342	117	265, 274	142	68, 107
93	3, 30, 66	118	265, 274	143	68, 107
94	35	119	342	144	68, 166
95	13, 39	120	281, 283, 284	145	70, 173
96	16, 30, 35	121	289	146	47, 165
97	16, 75	122	289	147	60
98	16, 70	123	/	148	213
99	70, 79	124	300, 315	149	206
100	30, 64, 101	125	321	150	206

Table 8 – Exercises and their Related Equations, Tables, and Chapters.

Ex.	Eq.	Ex.	Eq.	Ex.	Eq.
151	206, 207	176	314	201	382
152	206, 207	177	300	202	344, 376, 382
153	34, 66	178	333	203	C9
154	/	179	321, 322	204	C9
155	382	180	C7		
156	373, 375, 377	181	C7		
157	/	182	294		
158	390	183	247, C7		
159	224	184	247, C7		
160	213, 222	185	C9		
161	/	186	C9		
162	/	187	334, 327		
163	206, 213	188	333		
164	206, 213	189	300, 327, 333		
165	275, 277, 279	190	294, 333		
166	277, 279	191	300, 333		
167	15	192	393, 394, C9		
168	265, 270	193	393, 394, 395		
169	16, 279	194	47, 397, 398		
170	32, 259, 269	195	46, 400		
171	236	196	C8		
172	236, 277, 279	197	343, 345, 391		
173	61, 66	198	390		
174	283, 284	199	343, 345, 384		
175	281	200	373, 375, 377		

Table 9 – Exercises and their Related Equations, Tables, and Chapters.

PROBABILITY THEORY

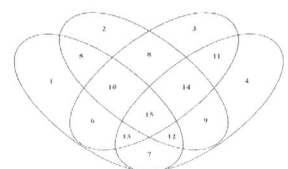

KEY CONCEPTS AND TOOLS
for
SOA EXAM P
& CAS EXAM 1

OLIVIER LE COURTOIS

www.ingramcontent.com/pod-product-compliance
Lightning Source LLC
Chambersburg PA
CBHW070234180526
45158CB00001BA/501